The Logica Yearbook 2015

The Logica Yearbook 2015

Edited by

Pavel Arazim

and

Michal Dančák

© Individual authors and College Publications 2016
All rights reserved.

ISBN 978-1-84890-213-8

College Publications
Scientific Director: Dov Gabbay
Managing Director: Jane Spurr

www.collegepublications.co.uk

Original cover design by Laraine Welch
Printed by Lightning Source, Milton Keynes, UK

All rights reserved. No part of this publication may be reproduced, stored in a retrieval system or transmitted in any form, or by any means, electronic, mechanical, photocopying, recording or otherwise without prior permission, in writing, from the publisher.

Preface

The book that you are holding in your hands is a further entry in a series of volumes that aspires to make some of the ideas presented at the annual international symposium Logica permanently accessible to both the conference participants and the wider public. The symposium, which took place at Hejnice Monastery in the Czech Republic from June 15 to June 19, 2015, brought together logicians from many different countries. This volume contains a representative sample of the contributions made at the conference.

The Logica symposium is an event with a tradition that began in the late nineteen eighties. Since that time, it has evolved into a respected conference which possesses a firm place in the annual schedule of the international community of logicians. Though the symposium is open to both researchers with a mathematical bent as well as a philosophical one, its audience traditionally consists mostly of logicians with philosophical interests. The informal atmosphere provides a space for a stimulating exchange of ideas among logicians of all generations, including students. As the editors of this volume we are proud that we can contribute to the successful completion of the annual symposium cycle by presenting this collection to you.

Last year's Logica was—as were all previous Logica symposia—organized by the Department of Logic of the Institute of Philosophy of the Czech Academy of Sciences. Twenty six lectures were presented during the conference, including those given by a distinguished list of invited speakers: Patricia Blanchette, Walter Carnielli, Melvin Fitting, and Peter Milne. As happens every year, the conference was enriched by a social programme that provided room for friendly debates concerning professional topics as well as for starting and developing personal friendships. The proceedings, which are traditionally published within one year of the conference, unfortunately offer only a very limited record of the topics discussed and cannot hope to even partially convey its atmosphere. In spite of that, we hope that you will find this book worthy of your attention.

Both the Logica symposium and The Logica Yearbook are the result of a joint effort by many people to whom we are very thankful. We would like to express our greatest gratitude to Petra Ivaničová, who had been a key member of the organizing crew for many years. Unfortunately she left the crew shortly after the symposium. We wish her good luck and success in her new career. We are, of course, very grateful to the Institute of Philosophy for all their support, without which the event wouldn't have been possible. We express our thanks to the staff of Hejnice Monastery for their hospitality and friendly assistance. Special thanks from the organizers and also, we believe, from the guests, go to the Bernard Family Brewery of Humpolec, which has traditionally sponsored the social programme of the symposium by providing three barrels of its excellent beer. We owe thanks to the Czech Science Foundation, which provided significant support for the meeting and for the publishing of this volume with the funding of the grant project no. 13-21076S. We would also like to thank College Publications and its managing director, Jane Spurr, for their very pleasant cooperation during the preparation of this book. Last, but not least, we would like to thank all of the authors for their exemplary collaboration during the editorial process.

Prague, April 2016

Pavel Arazim and Michal Dančák

Contents

Reference Graphs and Semantic Paradox 1
    Timo Beringer and Thomas Schindler

Models and Independence *circa* 1900 17
    Patricia Blanchette

Paraconsistency and Duality: between Ontological and
Epistemological Views 39
    Walter Carnielli and Abilio Rodrigues

From Diagrammatic to Mechanical Reasoning: the Case of
Syllogistic 57
    José Martín Castro-Manzano

Game Semantics for Vague Quantification 71
    Christian Fermüller

A Nonmonotonic Sequent Calculus for Inferentialist
Expressivists 87
    Ulf Hlobil

The Larger Logical Picture 107
    John Kearns

An Alternative Approach to Truth-value Semantics: *More or
Less True than* and Pairwise Valuations 117
    Rossella Marrano

Classical Logic Through the Looking-glass 133
    Peter Milne

Incompatibility and Inference as Bases of Logic 157
    Jaroslav Peregrin

Minimalism, Reference, and Paradoxes 163
    Lavinia Picollo

A Nonstandard Semantic Framework for Intuitionistic Logic 179
    Vít Punčochář

Internal Negation and Sortal Quantification 193
    Karel Šebela

Barbourian Temporal Logic 205
  Petr Švarný

On Paraconsistent Downward Löwenheim-Skolem Theorems 213
  Zach Weber

Reference Graphs and Semantic Paradox

TIMO BERINGER AND THOMAS SCHINDLER[1]

Abstract: We present a graph-theoretic analysis of the semantic paradoxes for the language of first-order Peano arithmetic augmented with a primitive truth predicate.

Keywords: Kripke's theory of truth, semantic paradox, semantic dependence, self-reference, games for truth, reference graphs

1 Introduction

The semantic paradoxes have traditionally been viewed as involving some vicious reference pattern such as self-reference. The paradigm of such a statement is the liar and it is plausible to represent the reference pattern underlying it by a simple loop:

L: (L) is false.

We can also consider pairs of sentences that, even if they are not directly self-referential, still exhibit some kind of circularity:

L_1: (L_2) is false; L_2: (L_1) is true.

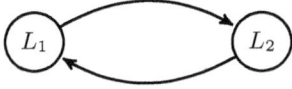

---

[1] This paper is an introduction to (Beringer & Schindler, 2016) to which we refer the interested reader for proofs of the theorems and many additional results. This work has been presented at conferences in Bristol, Buenos Aires, Hejnice, Helsinki, Ghent, Munich, and Oslo. We thank the audiences there for their valuable feedback. The work of the second author was generously supported by the Alexander von Humboldt Foundation.

Similarly, for every natural number n, we can consider liar cycles of length n. A slightly different example is given by the Curry paradox:

C_1: (C_1) is false or (C_2) is true; C_2 : $1+1=3$.

It is clear that self-reference or circularity is not a *sufficient* condition for paradox (consider e.g. 'This sentence contains five words.'). But is self-reference or circularity a *necessary* condition for paradox? Russell (1973) seems to assume as much, but the view has been challenged in recent decades. Herzberger (1970) argues that it is not circularity but rather groundlessness that makes a sentence pathological. Yablo (1993) also objects to Russell, drawing on the now famous example of an infinite sequence of sentences each which states that all subsequent sentences are false:

Y_1: (Y_n) is false for all $n > 1$; Y_2: (Y_n) is false for all $n > 2$; Y_3: (Y_n) is false for all $n > 3$; Y_4: (Y_n) is false for all $n > 4$; ...

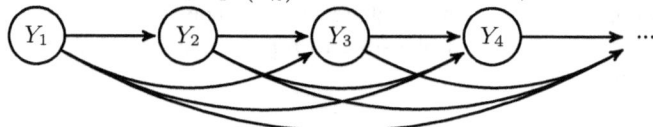

Variations on the Yablo sequence deliver new paradoxes whose reference graphs do not contain any loops. The following example is due to (Macauley, Rabern, & Rabern, 2013).

Y_1': (Y_2') is true; Y_2': (Y_n') is false for all odd $n > 2$; Y_3': (Y_4') is true; Y_4': (Y_n') is false for all odd $n > 4$; ...

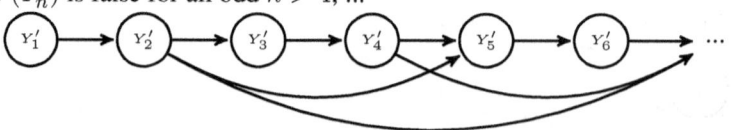

This raises the question how many types of paradox there are and what reference patterns underlie them. The idea of a graph-theoretic analysis of the semantic paradoxes is, of course, not new. We mention here particularly the work of Cook (2004) and Macauley et al. (2013). These authors work with infinitary *propositional* languages. Such languages offer a simple way of assigning a reference graph to every sentence of the language. Namely, sentence φ refers to sentence ψ if and only if there is a name α of ψ such that 'α is true (false)' is a syntactic constituent of φ. What will be done in the present paper is

- to provide, for the first time, a framework for quantificational first-order languages in which a reference graph is assigned to each sentence and a notion of *paradoxicality* can be defined in terms of possible decorations of a reference graph with truth values, in an analogous way as for propositional languages [section 2],

- to show that this notion of paradoxicality is co-extensional with that of Kripke (1975) with respect to a certain valuation scheme[2] [section 3, corollary 2],

- to provide a game-theoretic approach as a new tool for relating the reference pattern of a sentence to its paradoxicality [section 3],

- to conjecture an answer to the problem of characterizing the dangerous reference graphs [section 3, conjecture 1], in addition to providing a number of partial results [section 3, theorems 3-6],

- and to indicate how a more refined notion of reference can be formulated that allows to distinguish e.g. between the reference graph of the liar and that of the truth-teller. [section 4].

Technical preliminaries. The most straightforward way of formalizing the semantic paradoxes in first-order languages is by using the language of first-order Peano arithmetic augmented with a primitive unary predicate symbol T. We denote this language by $\mathcal{L}_T$. For each $\mathcal{L}_T$-sentence φ let $\#\varphi$ be the Gödel-number of φ with respect to some fixed Gödel-coding. $\mathcal{L}_T$ contains a name for each sentence φ—i.e., the numeral of $\#\varphi$—that we shall denote by $\ulcorner\varphi\urcorner$. By the (weak) diagonal lemma, we find for every $\varphi(x)$ a sentence ψ such that the biconditional $\psi \leftrightarrow \varphi(\ulcorner\psi\urcorner)$ is derivable from the axioms of Peano arithmetic. For example, letting $\varphi(x) := \neg Tx$ we obtain a liar sentence λ with $\lambda \leftrightarrow \neg T\ulcorner\lambda\urcorner$. In order to obtain a Yablo-sequence we need a generalized form of the diagonal lemma that allows for parameters. Then we get a predicate $Y(x)$ such that $\forall x[Y(x) \leftrightarrow \forall y > x \neg T\ulcorner Y(\dot{y})\urcorner]$

---

[2]This valuation scheme (V_L) is defined in section 2 and based on (Leitgeb, 2005). It is possible to extend this result to the Weak Kleene scheme (cf. section 4). The question to which other valuation schemes our framework is applicable must be left open for future research.

is provable in PA. Additionally, we can assume that the language contains function symbols for certain primitive recursive functions such as the substitution function. This allows us to formalize the semantic paradoxes in a more direct way. For example, we then obtain a term l such that the identity statement $l = \ulcorner \neg Tl \urcorner$ is provable in PA. If φ is a sentence and Φ a set of natural numbers, i.e., a set of $\mathcal{L}_T$-sentences (we shall often identify sentences with their codes) we let $Val_\Phi(\varphi)$ denote the truth value of φ in the classical interpretation $(\mathbb{N}, \Phi)$, where $\mathbb{N}$ is the standard model of arithmetic and Φ is the extension (interpretation) of the truth predicate T. We also write $(\mathbb{N}, \Phi) \models \varphi$ to indicate that φ is true in the model $(\mathbb{N}, \Phi)$.

2 Semantic dependence and reference graphs

In the introduction, we have assigned reference graphs to several paradoxical sentences relying merely on our intuitions about their referential relations.[3] In order to give a rigorous graph-theoretic analysis of the semantic paradoxes we need, of course, a systematic way of assigning a reference graph to every sentence of our language. In order to do so, we need a precise definition of what we may call, following Herzberger (1970), the *domain* of a sentence, i.e., the set of sentences that a given sentence refers to. Our proposal is to identify domains of sentences with dependence sets in the sense of Leitgeb (2005). Leitgeb defines the relation of semantic dependence between sentences and sets of sentences as follows:

Definition 1 *A sentence φ depends on $\Phi \subseteq \mathcal{L}_T$ iff for all $\Psi \subseteq \mathcal{L}_T$:* $Val_\Psi(\varphi) = Val_{\Phi \cap \Psi}(\varphi)$.

Thus, a sentence φ depends on a set of sentences Φ iff all sentences that are relevant for the evaluation of φ are contained in Φ. The operator $\mathbb{D}(\Phi) = \{\psi | \psi \text{ depends on } \Phi\}$ is monotonic: If $\Phi \subseteq \Psi$, then $\mathbb{D}(\Phi) \subseteq \mathbb{D}(\Psi)$. Let us call a set of sentences Φ $\mathbb{D}$-*sound* iff $\mathbb{D}(\Phi) \supseteq \Phi$.[4] Given any set of sentences S, we can iterate the operator $\mathbb{D}$ as follows: $\mathbb{D}_0(S) = S$ and $\mathbb{D}_\alpha(S) = \mathbb{D}(\bigcup_{\beta < \alpha} \mathbb{D}_\beta(S))$ for any ordinal $\alpha > 0$. If S is $\mathbb{D}$-sound then this process reaches a fixed point $G(S) = \bigcup_{\alpha \in On} \mathbb{D}_\alpha(S)$ and we call $G(S)$ the

---

[3]These intuitions, of course, shall prove to be accurate: According to the formal definition given in this section, each of the graphs depicted in the introduction is the canonical reference graph (of the $\mathcal{L}_T$-formalization) of the corresponding sentence.

[4]E.g., $\lambda_1 \notin \mathbb{D}(\{\lambda_1\})$ and $\lambda_2 \notin \mathbb{D}(\{\lambda_2\})$, where λ_1, λ_2 are the $\mathcal{L}_T$-versions of L_1, L_2. Hence neither of their singletons is $\mathbb{D}$-sound while $\{\lambda_1, \lambda_2\}$ is.

set of sentences *grounded in S*. A sentence is *grounded* (simpliciter) iff it is grounded in the empty set, and *ungrounded* otherwise.

By the monotonicity of the dependence operator, every sentence has infinitely many dependence sets. However, for certain sentences it is possible to single out a canonical dependence set. We say that a sentence φ depends *essentially* on a set Φ iff φ depends on Φ and there is no proper subset $\Psi \subset \Phi$ such that φ also depends on Ψ. This set Φ, if it exists, is unique. Most of the sentences usually considered in the literature on truth or the semantic paradoxes actually have essential dependence. However, there are many sentences which haven't. For example, consider the following version of the Yablo sequence which we may call the *nested* Yablo sequence:

$$Y^*(\overline{n}) \leftrightarrow \exists x > \overline{n} \forall y > x \neg \ulcorner T \ulcorner Y^*(\dot{y}) \urcorner \urcorner$$

The reader may verify that each $Y^*(\overline{n})$ lacks essential dependence: For all $m > n$, $Y^*(\overline{n})$ depends on $\{Y^*(\overline{m}), Y^*(\overline{m+1}), Y^*(\overline{m+2}), \ldots\}$ but does not depend on the intersection of these sets, the empty set; hence there is no least set on which $Y^*(\overline{n})$ depends.

Let us now see how we can relate Leitgeb's work to Kripke's theory of truth (Kripke, 1975). Therefore, consider the following monotonic valuation scheme.

Definition 2 *Let $S^+, S^- \subseteq \mathcal{L}_T$ be such that $S^+ \cap S^- = \emptyset$. We call the ordered pair (S^+, S^-) a partial model. The Leitgeb valuation scheme, V_L, is given by the following clauses:*

$$V_L(S^+, S^-)(\varphi) = \begin{cases} 1, & \text{if } \varphi \text{ depends on } S^+ \cup S^- \text{ and } (\mathbb{N}, S^+) \models \varphi \\ 0, & \text{if } \varphi \text{ depends on } S^+ \cup S^- \text{ and } (\mathbb{N}, S^+) \not\models \varphi \\ \frac{1}{2}, & \text{if } \varphi \text{ does not depend on } S^+ \cup S^-. \end{cases}$$

Obviously, V_L is a monotonic valuation scheme. Given (S^+, S^-), define the *Kripke-jump* $\mathcal{J}_L(S^+, S^-)$ as the ordered pair:

$$(\{\varphi | V_L(S^+, S^-)(\varphi) = 1\}, \{\varphi | V_L(S^+, S^-)(\varphi) = 0\})$$

The operator $\mathcal{J}_L$ is monotonic. Call a pair (S^+, S^-) *sound* iff $(S^+, S^-) \subseteq \mathcal{J}_L(S^+, S^-)$. A pair (S^+, S^-) is called a *fixed point* of $\mathcal{J}_L$ iff $\mathcal{J}_L(S^+, S^-) = (S^+, S^-)$. For any sound (X^+, X^-) there is a fixed point (S^+, S^-) of $\mathcal{J}_L$ with $(X^+, X^-) \subseteq (S^+, S^-)$ that can be obtained from (X^+, X^-) by iterating the jump operator $\mathcal{J}_L$ (possibly transfinitely). In particular, starting from the pair $(\emptyset, \emptyset)$ we reach the minimal fixed point. If (S^+, S^-) is a fixed point of $\mathcal{J}_L$ then for all $\varphi \in \mathcal{L}_T$ we have: $V_L(S^+, S^-)(\varphi) = 1 \Leftrightarrow \#\varphi \in S^+$ and $V_L(S^+, S^-)(\varphi) = 0 \Leftrightarrow \#\varphi \in S^-$. One easily proves that the minimal fixed point of $\mathcal{J}_L$ is identical to the fixed point model defined in (Leitgeb, 2005).[5] Thus, we have embedded Leitgeb's theory into Kripke's framework. This motivates the following definitions: A sentence φ is *Kripke-paradoxical* (with respect to V_L) iff there is no fixed point (S^+, S^-) of $\mathcal{J}_L$ such that φ has a definite truth value (i.e., 0 or 1) in (S^+, S^-); *Kripke-hypodoxical* (with respect to V_L) iff there are fixed points (S^+, S^-), (P^+, P^-) of $\mathcal{J}_L$ such that φ has a definite truth value in (S^+, S^-) and a different definite truth value in (P^+, P^-).

We are finally in a position to define the reference graphs for the sentences of $\mathcal{L}_T$. Let us call a function $f : \mathcal{L}_T \to \wp(\mathcal{L}_T)$ with $f(\varphi) \in \{\Phi | \varphi \text{ depends on } \Phi\}$ a *choice function*.

Definition 3 *The* reference graph *(short: rfg) G_φ of φ determined by f is defined as follows: The set $V(G_\varphi)$ of vertices (or nodes) of G_φ is the least set that contains φ and that contains with any ψ also all members of $f(\psi)$. Two vertices ψ, χ of G_φ are joined by an arc from ψ to χ iff $\chi \in f(\psi)$.*

Since every sentence has infinitely many dependence sets, every sentence has infinitely many rfgs. Some sentences φ, however, have a *canonical rfg*, i.e., an rfg of φ which is contained as a subgraph in every rfg of φ. Let us say that a sentence φ is *hereditarily essentially dependent* (hed) iff φ depends essentially on some set Φ and for every member ψ of Φ there is a Ψ such that ψ depends essentially on Ψ and so on. In particular, if a sentence has a finite rfg then that sentence is hed. It can be shown that a sentence has a canonical rfg iff it is hed.

Formally, an rfg G_φ can be seen as an attempt to recursively define the truth value of its root sentence φ. Call a function $d : V(G_\varphi) \to \{0, 1\}$ a *decoration* of G. A decoration is *acceptable* iff for all vertices ψ of G_φ

$$V_L(d_\psi^+, d_\psi^-)(\psi) = d(\psi),$$

where $d_\psi^+ = \{\chi \in out(\psi) | d(\chi) = 1\}$, $d_\psi^- = \{\chi \in out(\psi) | d(\chi) = 0\}$ and $out(\psi)$ is the set of all out-neighbours of ψ in G_φ. Acceptable decorations give us partial models validating the T-biconditionals of all sentences in $V(G)$ as follows:

---

[5] For an axiomatic theory of truth based on this model, see (Schindler, 2014).

Theorem 1 *Let d be an acceptable decoration of a reference graph G, $S_d^+ = \{\varphi | \varphi \in V(G) \wedge d(\varphi) = 1\}$ and $S_d^- = \{\varphi | \varphi \in V(G) \wedge d(\varphi) = 0\}$. Then for all $\varphi \in V(G)$: $V_L(S_d^+, S_d^-)(\varphi) = V_L(S_d^+, S_d^+)(T^\ulcorner\varphi\urcorner) \in \{0,1\}$.*

The following notions are from (Macauley et al., 2013) and adapted to our more general framework.[6] Let us call a sentence *r-paradoxical* ('r' for 'referentially') iff it has no rfg which admits an acceptable decoration; and *r-hypodoxical* iff it has an rfg which admits a verifying acceptable decoration and a falsifying acceptable decoration, where a decoration d of G_φ is *verifying* iff $d(\varphi) = 1$ and *falsifying* iff $d(\varphi) = 0$.[7]

For a sentence φ with a canonical rfg, the claim that φ is r-paradoxical (-hypodoxical) is equivalent to the claim that φ's canonical rfg does not admit an acceptable decoration (does admit a verifying and a falsifying decoration). Finally, call a graph *dangerous* iff it is isomorphic to an rfg of some r-paradoxical sentence. The problem of classifying all dangerous graphs is known as *the characterization problem*. Its solution is one of the most important goals of a graph-theoretic analysis of the paradoxes, and a mathematically challenging one.

3 Kripke-games on reference graphs

The main goals of the present section are (1) to show that G_φ has an acceptable decoration iff φ has a definite truth-value in some Kripke fixed point and (2) to establish criteria relating the existence of an acceptable decoration of G_φ to *structural properties* of G_φ, i.e., those properties which are invariant under isomorphism. To this end we will introduce a game (the verification game) being played on G_φ between two players ($\exists$) and ($\forall$) and link the existence of certain strategies for ($\exists$) to the existence of certain classes of acceptable decorations of G_φ. The verification game is parasitic on another game (the grounding game) which we discuss first.

---

[6]We slightly deviate from (Macauley et al., 2013). One reason is that they assign rfgs to *sentence systems*, not to single sentences, and define hypodoxicality (and paradoxicality) for sentence systems. In our terminology, a sentence system could be defined as the set of all vertices of an rfg. Another difference is that in their framework every sentence has a unique rfg. We discuss this in more detail in (Beringer & Schindler, 2016).

[7]In certain respects, our notion of Kripke-paradoxicality corresponds to the notion of inheritance-paradoxicality in (Yablo, 1982), and our notion of r-paradoxicality corresponds to what Yablo calls dependence-paradoxicality. Yablo doesn't have a notion corresponding to a reference graph but works with structures he calls dependence trees (our verification trees). How this concept is related to our reference graphs is briefly discussed in footnote 10. A more thorough discussion can be found in (Beringer & Schindler, 2016).

Timo Beringer and Thomas Schindler

The grounding game

The rules of $\mathcal{G}_G(\varphi, \Phi)$ are the following:

- ($\forall$) must move first and choose φ as his first move φ_1.
- As her n-th move ($\exists$) must choose some set Φ_n on which φ_n depends.
- If $n > 1$, as his n-th move ($\forall$) must choose some sentence $\varphi_n \in \Phi_{n-1} \setminus \Phi$.

The winning conditions for $\mathcal{G}_G(\varphi, \Phi)$ are: ($\exists$) wins a run of the game if ($\forall$) cannot move. ($\forall$) wins a run of the game if it goes on forever.[8]

We have a special interest in cases where the set parameter Φ denotes the empty set; we then omit the parameter and write $\mathcal{G}_G(\varphi)$. A *strategy for* ($\exists$) *in* $\mathcal{G}_G(\varphi)$ is a non-empty set σ of possible positions such that (i) whenever $(\varphi_1, \Phi_1, \ldots, \Phi_n) \in \sigma$ then $(\varphi_1, \Phi_1, \ldots, \Phi_n, \psi) \in \sigma$ for all $\psi \in \Phi_n$, and (ii) whenever $(\varphi_1, \Phi_1, \ldots, \Phi_{n-1}, \varphi_n) \in \sigma$ then $(\varphi_1, \Phi_1, \ldots, \Phi_{n-1}, \varphi_n, \Phi_n) \in \sigma$ for a unique set Φ_n on which φ_n depends. Thus, a strategy for ($\exists$) is a tree with φ as its root, and each branch is a possible run of the game. Analogously a strategy for ($\forall$) can be defined. Call an ($\exists$)-strategy *homogeneous* iff for each sentence ψ holds: if ($\exists$) plays Ψ in response to some ($\forall$)-move ψ in some σ-position, then ($\exists$) plays Ψ in response to ψ in all σ-positions.

A notion closely related to that of an ($\exists$)-strategy is that of a *dependence tree*: A dependence tree T is a tree such that each node of T depends on the set of its T-children. There is a canonical bijection between ($\exists$)-strategies and dependence trees: delete all the set-components (the Φ_n's) from all positions of σ. The result is a dependence tree T_σ, which can be returned into σ by recursively attaching to each sequence the set of the T_σ-children of its last element. The image of a homogeneous strategy under this bijection is a homogeneous dependence tree. A homogeneous dependence tree, in turn, can be associated to a (partial) choice function by selecting for each sentence occurring in it the set of its children as dependence set. Each choice function, on the other hand, determines, together with a distinguished sentence φ, a dependence tree with φ as its root. Since any rfg can be *unfolded*[9] into

---

[8] In (Welch, 2009) an extensive collection of games can be found, each of which allows the characterization of the T-predicate (suggested by a particular formal theory of truth) in terms of a player's strategies. For our purposes, however, a mere characterization of the T-predicate's extension is not enough: we need a transparent reconstruction of the valuation process leading to this extension in the rules of the game.

[9] An rfg G of φ is a *pointed accessible graph* in the sense of (Aczel, 1980): 'pointed' means that it has a distinguished node, its *root* φ, while 'accessible' means that each of its nodes ψ

a homogeneous dependence tree, it corresponds canonically to a homogeneous ($\exists$)-strategy. On the other hand, every ($\exists$)-strategy σ can be *collapsed* into an rfg $\Gamma(\sigma)$ whose unfolding is the dependence tree T_σ: The set of vertices of $\Gamma(\sigma)$ consists of the sentences occurring in σ; two vertices ψ, χ are joined by an arc from ψ to χ iff $(\varphi, \ldots, \psi, \Psi, \chi) \in \sigma$ for some Ψ, i.e., if there is a run of the game (played according to σ) in which ($\forall$) chooses ψ, χ consecutively. As a consequence, we can interpret the grounding game for a sentence φ as being *played on an rfg of* φ, namely the rfg that ($\exists$)'s chosen strategy collapses to.

The grounding game derives its name from the fact that given any $\mathbb{D}$-sound set of sentences S, φ is grounded in S iff ($\exists$) has a winning strategy in the game $\mathcal{G}_G(\varphi, S)$. Notice that a winning strategy for ($\exists$) is a well-founded tree. We therefore obtain that a strategy σ for ($\exists$) in $\mathcal{G}_G(\varphi)$ is a winning strategy for ($\exists$) iff $\Gamma(\sigma)$ is well-founded. Combining the last two results, we obtain that a sentence is grounded iff it has a well-founded rfg.

The verification game

The *verification game* $\mathcal{G}_T(\varphi, v, \mathcal{F})$ is quite similar to the grounding game $\mathcal{G}_G(\varphi, \Phi)$, but this time the players are not dealing merely with sentences φ and sets of sentences Φ, but with facts and sets of facts $\mathcal{F}$. A *fact* is an ordered pair (φ, v), consisting of a sentence φ and a truth value v that can be either 0 or 1. (This notion of fact was introduced by Yablo (1982).) We let $\mathcal{F}^+ = \{\varphi | (\varphi, 1) \in \mathcal{F}\}$ and $\mathcal{F}^- = \{\varphi | (\varphi, 0) \in \mathcal{F}\}$. Thus, sets of facts are partial interpretations of the truth predicate (an extension and an anti-extension) encoded as a single set. Therefore, we often identify sets of facts and partial models. We say, for instance, that $\mathcal{F}$ is a sound set of facts, meaning that $\mathcal{F}$ considered as the partial model $(\mathcal{F}^+, \mathcal{F}^-)$ is sound in the sense of Kripke. Let $\|(\varphi, v)\| := \varphi$ and $\|\mathcal{F}\| := \|(\mathcal{F}^+, \mathcal{F}^-)\| := \mathcal{F}^+ \cup \mathcal{F}^-$. (Sets of) facts can be seen as decorations of (sets of) sentences with truth values: a fact (φ, v) is a *a decoration* of ψ iff $\|(\varphi, v)\| = \psi$ and a set of facts $\mathcal{F}$ is a *a decoration* of Φ iff $\|\mathcal{F}\| = \Phi$. A second difference to the grounding game is that a run of the verification game can end in a draw.

To every position of the game $\mathcal{G}_T(\varphi, v, \mathcal{F})$ a *mode* is associated, the mode that a run of the game assumes in this position. This mode is either the *verification mode* or the *falsification mode*.

---

can be reached in a *walk* starting from the root φ, i.e., there is a sequence $\varphi_1, \ldots, \varphi_n$ of nodes such that $\varphi_1 = \varphi$, $\varphi_n = \psi$ and for each $1 \leq i < n$ there is an arc of G from φ_i to φ_{i+1}. The unfolding of graph G is the tree consisting of all finite walks in G starting from its root.

The rules of $\mathcal{G}_T(\varphi, v, \mathcal{F})$ are:

- The game $\mathcal{G}_T(\varphi, 1, \mathcal{F})$ starts in the verification mode, $\mathcal{G}_T(\varphi, 0, \mathcal{F})$ starts in the falsification mode.
- ($\forall$) must move first and choose φ as his first move φ_1.
- As her n-th move, ($\exists$) must choose some partial model (Φ_n^+, Φ_n^-) such that φ_n depends on $\Phi_n^+ \cup \Phi_n^-$ and $Val_{\Phi_n^+}(\varphi_n) = 1$ if the game is in verification mode, and $Val_{\Phi_n^+}(\varphi_n) = 0$ if the game is in falsification mode.
- If $n > 1$, as his n-th move ($\forall$) must choose some sentence $\varphi_n \in (\Phi_{n-1}^+ \setminus \mathcal{F}^+) \cup (\Phi_{n-1}^- \setminus \mathcal{F}^-)$. If $\varphi_n \in \Phi_{n-1}^+$ then play continues in the verification mode. If $\varphi_n \in \Phi_{n-1}^-$ then play continues in the falsification mode.

The winning condition for $\mathcal{G}_T(\varphi, v, \mathcal{F})$: If a player cannot move according to the rules, then the other player wins this run of the game. If a run of the game goes on forever it is declared a draw.

In order to relate our two notions of paradoxicality (cf. thms 3 and 4), special attention is payed to cases where the set parameter $\mathcal{F}$ denotes the empty set; we then write $\mathcal{G}_T(\varphi, v)$. Strategies for the verification game are similar to strategies for the grounding game with one important difference. Since we want to keep track of the mode of game, we represent the possible positions as follows: $((\varphi_1, v_1), (\Phi_1^+, \Phi_1^-), \ldots, (\varphi_n, v_n), (\Phi_n^+, \Phi_n^-))$, where v_i is either 1 or 0 according as to whether the game is in verification or falsification mode after ($\forall$)'s i-th move. (Thus a position is an alternating sequence of facts and partial models.) The verification game derives its name from the property that φ is true in the fixed point of $\mathcal{J}_L$ generated by $\mathcal{F}$ iff ($\exists$) has a winning strategy in $\mathcal{G}_T(\varphi, 1, \mathcal{F})$ and φ is false in this fixed point iff ($\exists$) has a winning strategy in $\mathcal{G}_T(\varphi, 0, \mathcal{F})$.

Let $\|\sigma\|$ be the set of positions which arises from an ($\exists$)-strategy σ by replacing each position $p = ((\varphi_1, v_1), (\Phi_1^+, \Phi_1^-), \ldots, (\varphi_n, v_n), (\Phi_n^+, \Phi_n^-))$ of σ by $\|p\| = (\|(\varphi_1, v_1)\|, \|(\Phi_1^+, \Phi_1^-)\|, \ldots, \|(\varphi_n, v_n)\|, \|(\Phi_n^+, \Phi_n^-)\|)$. We say that σ is *a decoration* of $\|\sigma\|$. Clearly $\|\sigma\|$ is an ($\exists$)-strategy in the grounding game. Thus, a strategy in the verification game can be seen as the result of decorating a strategy in the grounding game with truth values. Of course, there are many ways of decorating a grounding strategy. This changes under the stipulation that winning strategies must be mapped to winning strategies:

Reference Graphs and Semantic Paradox

Theorem 2 *For each winning* ($\exists$)-*strategy* σ *in* $\mathcal{G}_G(\varphi, \Phi)$ *and each decoration* $\mathcal{F}$ *of* Φ *there is a unique truth value* v *and a unique winning* ($\exists$)-*strategy* σ^* *in* $\mathcal{G}_T(\varphi, v, \mathcal{F})$ *such that* σ^* *is a decoration of* σ.

Call a verification-strategy σ *homogeneous* iff the underlying strategy $\|\sigma\|$ is homogeneous. Given the projection $\|\cdot\|$ and the relation of strategies in the grounding game to reference graphs we can also interpret the verification game for a sentence φ as being *played on an rfg of* φ, namely the rfg G_φ to which ($\exists$)'s underlying strategy collapses. In this way any ($\exists$)-strategy σ induces a *multi-decoration* D_σ on G_φ, i.e., a function $D_\sigma : V(G_\varphi) \to \{0, 1, \{0, 1\}\}$. Such a decoration D_σ may be *consistent* or *inconsistent* according to whether each node is assigned only one truth value or two. Any consistent multi-decoration is a decoration of G_φ.[10]

Now let us investigate how acceptable decorations of rfgs are related to strategies in the verification game. In order to do so, we cannot focus solely on winning strategies but need a somewhat more liberal criterion for a good ($\exists$)-strategy σ in the verification-game: Call σ *faithful* iff ($\exists$) never loses a game whenever she plays σ and call σ *consistent* iff no sentence occurring in σ occurs in both the verification- and the falsification-mode.

Theorem 3 *Let G be an rfg of φ and d be a multi-decoration of G. Then d is an acceptable verifying decoration of G iff it is induced by a faithful consistent (and homogeneous)*[11] *($\exists$)-strategy in $\mathcal{G}_T(\varphi, 1)$, and d is an acceptable falsifying decoration of G iff it is induced by a faithful consistent (and homogeneous) ($\exists$)-strategy in $\mathcal{G}_T(\varphi, 0)$.*

Paradoxicality and a graph's structural properties

According to Yablo (1982), the paradoxicality of a sentence lies in the fact that "when we unravel and chase down the sentences truth *or* falsity conditions, we are led to something absurd. And absurd here can only mean one of two things: either we are led to call a true (false) sentence false (true) (as when, for example, we choose to deny that Epimenides was really a Cretan),

---

[10] Just as an ($\exists$)-strategy in the grounding game corresponds to a dependence tree, an ($\exists$)-strategy in the verification game corresponds to a *verification tree*, i.e., to a dependence tree decorated with truth values in such a way that $V_L(S_\psi^+, S_\psi^-)(\psi) = v(\psi)$ for any node ψ, where S_ψ^+ is the set of ψ's children decorated with 1 and S_ψ^- is the set of ψ's children decorated with 0. It is easily seen that our notion of a verification tree coincides with Yablo's notion of a dependence tree (Yablo, 1982) if the Strong Kleene Valuation scheme in his definition is replaced by V_L.

[11] Here and in the following theorems, the condition of homogeneity can be dropped.

or we are led to maintain of a sentence that it is both true and false (as when we concede Epimenides nationality and elect to wrestle with the resulting self-dependency of his utterance)." The following theorem can be seen as a precise formulation of that claim.[12]

Theorem 4 *A sentence φ has the truth value v in some Kripke fixed point (Φ^+, Φ^-) iff $(\exists)$ has a faithful consistent (and homogeneous) strategy σ in $\mathcal{G}_T(\varphi, v)$. Moreover $(S^+, S^-) \subseteq (\Phi^+, \Phi^-)$, where (S^+, S^-) is the decoration induced by σ on the rfg that σ collapses to.*

Corollary 1 *Every well-founded rfg admits a unique acceptable decoration.*

Corollary 2 *A sentence is Kripke-paradoxical iff it is r-paradoxical.*

Kripke (1975) writes: "The largest intrinsic fixed point is the unique 'largest' interpretation of $T(x)$ which is consistent with our intuitive idea of truth and makes no arbitrary choices in truth assignments." The following theorem formulates this 'non-arbitrariness' in terms of decorations:

Corollary 3 *A sentence is in the largest intrinsic fixed point of $\mathcal{J}_L$ iff it has an rfg which admits either a verifying acceptable decoration or a falsifying acceptable decoration (but not both). As a consequence, a sentence is Kripke-hypodoxical iff it is r-hypodoxical.*

The following theorem shows that we can always choose the second possibility mentioned in the above quote from Yablo's paper.

Theorem 5 *Every rfg admits a multi-decoration which is induced by a faithful homogeneous $(\exists)$-strategy in the verification-game.*

Aside from its rather philosophical meaning hinted at above, theorem 5 allows us (together with theorem 3) to identify certain structural properties that all rfgs of a sentence share as necessary condition for its paradoxicality: Since there is always a faithful decoration of any rfg G_φ, the paradoxicality of φ must be due to the fact that all of the faithful decorations of all of φ's rfgs are lacking consistency. But the property of lacking a consistent decoration can be related to a graph's structural properties rather easily:

Corollary 4 *If a sentence φ has a reference graph which is a tree, then φ is not r-paradoxical.*

---

[12]Moreover, it can be seen as a reformulation of the last theorem in (Yablo, 1982) in our framework.

If an rfg is not a tree then it either contains a directed cycle as a subgraph or it contains a type of subgraph that we may call a *double path*, i.e., a graph consisting of two paths originating both from the same vertex and rejoining in a different vertex, not touching each other in between. The reason for this is that rfgs can be seen as collapsed dependence trees (which in turn are their unfoldings).

Corollary 5 *If a sentence is r-paradoxical, then each of its rfgs contains a directed cycle or a double path.*

Now let us turn to the classification problem for dangerous graphs. First, we can formulate and prove in our framework the following result of (Macauley et al., 2013):

Theorem 6 *A finite rfg is dangerous iff it contains a directed cycle.*

It is worth noticing that while the directed cycle is the reference pattern underlying the liar family, the double path is underlying any member of the Yablo sequence. However, it can be shown that if φ has an rfg with no cycles and only *finitely* many double paths, then φ is not r-paradoxical. Unlike cycles, double paths must come in flocks in order to make an rfg dangerous.

Conjecture 1 *A graph is dangerous iff it is isomorphic to some rfg of a sentence and contains a subdivision of the liar-graph as a subgraph or the Yablo-graph as a finitary minor.*[13]

This conjecture is motivated by an attempt to make the notion of a graph containing *many* double paths precise: An acyclic graph should contain the Yablo-graph as a finitary minor iff it deviates considerably from being a tree in the sense that it contains many double paths. This conjecture, if correct, implies that in some sense every r-paradoxical sentence is reducible either to the liar or the Yablo paradox.

4 Outlook

It should be obvious that paradox cannot be characterized purely in referential terms. For example, the canonical rfg of the liar and the truth-teller

---

[13]For the notion of a minor in the context of digraphs consult the appendix of (Beringer & Schindler, 2016). Intuitively, the minor-relation is a more liberal form of the subgraph-relation that allows that connected sets of vertices can be contracted. By *finitary* we mean that any set of vertices which is contracted to one vertex must be finite.

are isomorphic, both consist merely of a simple loop. But while the liar is paradoxical, the truth-teller is not. Intuitively, what distinguishes both sentences is that the first makes a 'negative' statement about itself, claiming itself to be false, while the second makes a 'positive' statement about itself, claiming itself to be true. Accordingly, we should label the rfg of the liar with a '+' and the rfg of the truth-teller with a '−'. The verification game provides us with the means of labelling every rfg in a systematic manner. Recall that for every rfg there is a strategy σ in the verification game such that there is an arc from φ to ψ iff there is a run of the game (following that strategy) in which φ, ψ are chosen consecutively by ($\forall$); call such a pair a *transition*. Recall furthermore that to every position in the verification game there corresponds a mode, the verification or falsification mode. Now label the arc from φ to ψ with a plus if there is a transition from φ to ψ with no change in the mode, and a minus if there is a transition from φ to ψ with a change in the mode. This definition depends on the choice of σ and does not exclude the possibility of incoherent arcs, i.e., arcs labelled with both a plus and a minus. Let us call rfgs that result from this process of labelling *signed* rfgs. Signed rfgs provide us with a more fine-grained picture of the paradoxes. We mention the following result: If a sentence φ has a faithful signed rfg G_-^+ such that there is no cycle in G_-^+ which contains an incoherent arc or an odd number of negative arcs and there is no double path in G_-^+ which contains an incoherent arc or both branches of which have a different number modulo 2 of negative arcs, then φ is not r-paradoxical.

In concluding this paper, let us make a brief remark on the paper of Macauley et al. (2013). As mentioned in the introduction, Macauley et al. work with an infinitary propositional language and the notion of reference operative there is essentially *reference by mentioning*. This notion of reference is strongly compositional and one can interpret this compositionality straightforwardly as Weak Kleene dependence. It is then easy to show that the notion of paradoxicality used by Macauley et al. is co-extensional with the notion of Kripke-paradoxicality with respect to the Weak Kleene scheme formulated for their propositional language. In particular, it allows a reconstruction of the framework of Macauley et al. within our own. This raises the question to which valuation schemes other than V_L or Weak Kleene our framework can possibly be expanded. We hope to be able to give an answer in our future work.

References

Aczel, P. (1980). *Non-well-founded Sets*. Center for the Study of Language and Information.

Beringer, T., & Schindler, T. (2016). A Graph-theoretic Analysis of the Semantic Paradoxes. *In preparation*.

Cook, R. T. (2004). Patterns of Paradox. *Journal of Symbolic Logic, 69*(3), 767–774.

Herzberger, H. (1970). Paradoxes of Grounding in Semantics. *Journal of Philosophy, 67*, 145–167.

Kripke, S. (1975). Outline of a Theory of Truth. *Journal of Philosophy, 72*, 690–716.

Leitgeb, H. (2005). What Truth Depends on. *Journal of Philosophical Logic, 34*(2), 155–192.

Macauley, M., Rabern, B., & Rabern, L. (2013). Dangerous Reference Graphs and Semantic Paradoxes. *Journal of Philosophical Logic, 42*(5), 727–765.

Russell, B. (1973). On some Difficulties in the Theory of Transfinite Numbers and Order Types. In *Essays in analysis* (pp. 135–164). London: Allen and Unwin.

Schindler, T. (2014). Axioms for Grounded Truth. *Review of Symbolic Logic, 7*, 73–83.

Welch, P. (2009). Games for Truth. *Bulletin of Symbolic Logic, 15*(4), 410–427.

Yablo, S. (1982). Grounding, Dependence, and Paradox. *Journal of Philosophical Logic, 11*(1), 117–137.

Yablo, S. (1993). Paradox without Self-reference. *Analysis, 53*, 251–252.

Timo Beringer
LMU Munich
Germany
E-mail: timob2001@yahoo.de

Thomas Schindler
Clare College, University of Cambridge
United Kingdom
E-mail: ts647@cam.ac.uk

Models and Independence
circa 1900

PATRICIA BLANCHETTE[1]

Abstract: The half-century from 1870 to 1920 involved the intrusion into logic of the notion of model in ways that altered both the palette of tools available to the logician, and the conception of the logical properties and relations that form the subject-matter of logic. Prior to 1880, the central role of models in mathematics was a purely geometric one: models were geometric constructions used to provide a substrate for non-Euclidean geometry, and subsequently to demonstrate the independence of the parallels postulate. But by 1900, some dramatic changes had taken place, with models now consisting of arbitrary interpretations of mathematical vocabulary, with broad application outside of the parallels postulate, and outside of geometry, to the demonstration of consistency and independence results throughout mathematics. And by 1918, the technique of using models to prove independence was applied to logic itself, in a yet-broader conception of "model," to prove the independence of fundamental logical principles one from another. This essay investigates the broadening of the conception of model during this period, and a closely-connected change in the conception of independence and associated notions. We will be particularly interested in the ways in which changes in the notion of model have gone hand-in-hand with changes in our understanding of the role of axioms in mathematical theories. Especially important will be an investigation of the ways in which our contemporary understanding of independence, consistency, and logical entailment have been influenced by the growing role of models in the analysis of these notions.

Keywords: models, independence, history of logic, geometry, consistency

---

[1] Parts of this material were presented to the Philosophy Department at the Ohio State University, and to the 2014 meeting of the Society for the Study of the History of Analytic Philosophy in Montreal; thanks to audience members on both occasions for helpful discussion. Thanks also to the organizers of Logica 2015 for an excellent conference, and for helpful discussion of this material. Thanks to the Institute for Scholarship in the Liberal Arts at the University of Notre Dame for financial assistance.

Patricia Blanchette

1 Introduction

Questions concerning the *independence* of specific propositions from collections of other propositions have been central to logic and mathematics for as long as our history of these disciplines records. Taking independence to be the denial of logical entailment, the centrality of this notion is immediate: to propose that a given proposition is independent of others is just to propose that it is not (in some sense) entailed by those others, so that a concern with logical entailment is at the same time a concern with independence. Similarly, a concern with the *consistency* of a given theory is a concern with independence, as long as we understand a theory to be consistent iff no contradiction is entailed by it. But despite the centrality of the notion (or, perhaps, of the family of related notions) of independence, the idea that we might *prove* a given proposition to be independent of others was, until the nineteenth century, a daunting idea. The independence of the parallels postulate, for example, is generally taken to have been decisively established as late as 1868 with Beltrami's pseudosphere, a Euclidean construction of a "model" of a two-dimensional non-Euclidean geometry.[2] It took another thirty years for the model-construction technique to be refined and applied broadly outside of geometry, and at least another decade for the refinement to enable independence proofs in logic itself.

The straightforward story that suggests itself is that the development just sketched is essentially the construction and subsequent refinement of the tools necessary to decisively treat, by the dawn of the twentieth century, those questions of independence that had already for centuries been considered, studied, and conjectured, but never rigorously answered.[3]

In what follows, I will suggest that this is not the right way to understand the history of our involvement with independence questions. What I will suggest is that we have not, over the last few centuries, been engaged with a single relation of independence in whose demonstration we have become increasingly skilled. Instead, as our mathematical and logical tools have changed, so too have the independence questions that we are in a position to ask, and to answer. One result of this is that despite our modern sophistication, it is not at all clear that the techniques we now have at our disposal, those involving the modern notion of "model," are sufficient totreat

---
[2](Beltrami, 1868).
[3]For such a characterization of the development, see e.g. (Nagel, 1939).

Models and Independence *circa* 1900

the questions that, in previous eras, would naturally have been taken to be the central independence questions.

To see how the techniques for proving independence, and the focal independence questions themselves, evolved over the period of roughly the half-century surrounding 1900, we will look here at just a few highlights of that development. We begin in the middle of the story, at the turn of the nineteenth to the twentieth century, when modern "model-theoretic" techniques have begun to take recognizable form. We then backtrack to consider some important antecedents of those techniques, and finally move forward into the even more-modern era to see how the further rigorization of method in the early twentieth century brought about a further shift in the kinds of independence questions that came to occupy center stage. My hope is that greater clarity about the evolution of our views and techniques concerning independence during this critical period will bring with it greater clarity about what our methods do in fact show, and greater clarity about the various relations of independence themselves.

2 Hilbert 1899

We begin by looking at the state of the art of independence proofs in the penultimate year of the nineteenth century. Here our example is David Hilbert's 1899 *Foundations of Geometry*, in which we find a clear and systematic application of the technique that was becoming, at this point, the standard approach to demonstrations of independence. As Felix Klein puts it in 1908, describing what he calls the "modern theory" of geometric axioms:

> In it, we determine what parts of geometry can be set up without using certain axioms ...
> As the most important work belonging here, I should mention Hilbert's [1899].[4]

Hilbert's work in the *Foundations of Geometry* monograph involves a clear and careful axiomatization of Euclidean geometry, together with a consistency proof for the whole, and a series of independence proofs that demonstrate the connections of logical entailment holding between the various parts of the edifice. Hilbert characterizes both consistency and independence here in terms of the relation of logical deducibility: a set of axioms is

---

[4](Klein, 1908), as quoted in (Bennett & Birkhoff, 1988, p. 185).

consistent, he tells us, if "[I]t is impossible to deduce from them by logical inference a result that contradicts one of them" [§9], and a geometric axiom or theorem is *independent* of a collection thereof if it cannot be deduced from that collection.

Hilbert's method of demonstrating non-deducibility is as follows: Given a set AX of axioms, and a statement (perhaps another axiom) A, we begin by uniformly re-interpreting the geometric terms ("point," "line," "lies-on," etc.) in AX and in A, in terms of objects and relations given by a different theory, in this case a theory R of real numbers and collections thereof. We then note that, as re-interpreted, each of the sentences AX, together with any sentence deducible from them, expresses a theorem of R. Finally, the negation $\neg A$ of the target sentence A also expresses a theorem of R, which, assuming the consistency of R, guarantees that A itself does not express a theorem of R. Still assuming the consistency of R, then, we have a guarantee that A is not deducible from AX.

The consistency of AX, in the sense of the non-deducibility of a contradiction from AX, is demonstrable similarly, again assuming the consistency of the background theory R. As Hilbert says,

> From these considerations, it follows that every contradiction resulting from our system of axioms must also appear in the arithmetic related to the domain [of the background theory].[5]

An important point to note about the interpretation-theoretic technique used by Hilbert here is that it presupposes that the relation of deducibility in question is "formal" in the sense that it is unaffected by the reinterpretation of geometric terms. It is this that guarantees that the sentences deducible from AX will express theorems of R under the reinterpretation, given only that the members of AX do. But the deducibility relation is not, for Hilbert in 1899, "formal" in the sense of "syntactically specified;" there is no formal language at this point, and no explicit specification of logical principles. We will use the term "semi-formal" for such a relation. The first thing, then, that Hilbert's interpretations (or models) shows is that a given sentence is not semi-formally deducible from a collection of sentences.

Hilbert's models also show, importantly, the *satisfiability* of the conditions implicitly defined by the collections of sentences in question. Given a collection $AX \cup \neg A$ of sentences whose geometric terms appear schematically, a Hilbert-style reinterpretation on which each member of that collec-

---

[5](Hilbert, 1899, §9).

tion expresses a truth about constructions on the real numbers demonstrates the satisfiability of the condition defined by the collection. Equivalently, it demonstrates that the condition defined by AX can be satisfied without satisfying the condition defined by A.

Hilbert's technique, then, demonstrates the *independence* of a given sentence from a collection of sentences in two different senses. Taking as an example the question of the independence of Euclid's parallel postulate (PP) from the remainder of the Euclidean axioms (E) for the plane, the two senses, with our labels introduced, are:

- $Independence_D$: (PP) is not (semi-formally) deducible from (E);

- $Independence_S$: The condition defined by $(AX) \cup \neg(PP)$ is satisfiable.

$Independence_S$ is the stronger of the two notions, though they are extensionally equivalent in the setting of an ordinary first-order language.[6]

3 A third kind of independence

Gottlob Frege's work, in the same period, focuses on a notion of independence that's distinct from both of the relations demonstrable via Hilbert's technique. For Frege, *independence* is a relation not between sentences but between *thoughts*, i.e. between the kinds of things expressible by fully-interpreted sentences. Each thought, as Frege understands it, has a determinate subject-matter: thoughts about geometric objects and relations are entirely distinct from thoughts about collections of real numbers. Hence the re-interpretation of sentences along Hilbert's lines will result in the assignment to those sentences of different thoughts. Finally, logical connections between thoughts, connections like dependence and independence, provability and consistency, are sensitive as Frege sees it to the contents of the simple terms in the sentences used to express those thoughts. Hence Hilbert's re-interpretation strategy amounts, from Frege's point of view, to shifting attention from the geometric thoughts in which one was originally interested to an entirely different collection of thoughts, a collection whose

---

[6]The equivalence is given by the completeness of first-order logic. Hilbert's setting is that of natural language without strictly-defined relations of deducibility or satisfiability, so the question of the extensional relationship between the two independence relations is imprecise. The expressive richness of that language, however, is well beyond that of (what was to become) first-order logic, giving the second relation, in that setting, a narrower extension than the first.

logical properties are no guide to those of the original target thoughts. As a result, Frege takes it that Hilbert's technique is unsuccessful in demonstrating consistency and independence.

> Mr. Hilbert appears to transfer the independence putatively proved of his pseudo-axioms to the axioms proper. ... This would seem to constitute a considerable fallacy. And all mathematicians who think that Mr. Hilbert has proved the independence of the real axioms from one another have surely fallen into the same error.[7]

Or, as we might more calmly put it, the relation that Frege calls "independence" is neither of the relations, also reasonably known by that name, demonstrable via Hilbert's technique. For Frege, the parallels postulate is independent of the other axioms of Euclid iff it isn't provable from those axioms (and in this he agrees with Hilbert), but the notion of *proof* that Frege works with is a very rich one: the question of whether a given thought is provable from others can turn on non-trivial conceptual analyses of the components of those thoughts.[8] Hence a sentence A can be Independent$_D$ and even Independent$_S$ of a set AX of sentences while the thought $\tau(A)$ expressed by A fails to be independent in Frege's sense of the set $\tau(AX)$ of thoughts expressed by the members of AX. We introduce a term for this third kind of independence:

- *Frege − independent*: The thought $\tau(A)$ is not provable, in Frege's rich sense, from the set $\tau(AX)$ of thoughts.

To fix ideas with a vivid example: consider the sentences

(1) Point B lies on a line between points A and C.

(2) Point B lies on a line between points C and A.

For Hilbert, a model can immediately show that (2) is independent of (1), in both of the relevant senses: (2) is not semi-formally deducible from (1),

---

[7](Frege, 1906, p. 402).

[8]Rigorous deduction, for Frege as for Hilbert, cannot make reference to the meanings of non-logical terms. But the demonstration that a given thought τ is provable from a collection Σ of thoughts can (and, in the logicist project, clearly does) involve non-trivial analysis of τ and/or of Σ *en route* to the expression of those thoughts in the sentences that will appear in the rigorous deduction. See (Blanchette, 1996, 2012).

Models and Independence *circa* 1900

and the condition defined by (1) ∪ ¬(2) is satisfiable.[9] For Frege on the other hand, a model can show no such thing. Though Frege does not discuss this example, it is compatible with his views that the thoughts expressed by sentences (1) and (2) are the same thought, and hence provable immediately from one another.

4 The parallels postulate

The technique of reinterpreting the language to show independence in Hilbert's way was new in the decade leading up to 1900. But the question of the independence of the parallels postulate was by then centuries old, and the conviction that it had an affirmative answer was, before the interpretation-theoretic technique came on the scene, well entrenched. This makes pressing the following questions: What was meant by "independence" prior to this period? And what was it that convinced mathematicians, prior to the 1890's, of the independence of the parallels postulate?

The following discussion is not intended as a complete answer to these questions. The goal is rather to examine a few significant points along the route to the modern era of independence proofs, as a means of clarifying some of the conceptual changes that were necessary to the progress along that route.

We begin with a brief mention of J. H. Lambert (1728–1777). Lambert famously examined the independence question by working out some of the fundamental implications of $(\neg PP)$, the negation of the parallels postulate. He noted, importantly, that $(\neg PP)$ holds of spherical triangles. But he did not take this fact to support the independence hypothesis. On the contrary, despite recognizing the "non-Euclidean" (as *we* might put it) behavior of the sides of spherical triangles, Lambert continued to look for a proof of the parallels postulate from the rest of Euclid.

In answer to the question why Lambert didn't take the example of the spherical triangles as support for the independence thesis, Katherine Dunlop writes, of Lambert's view of the question "whether the principles that hold of figures on [the surface of a sphere] constitute a theory that is genuinely comparable to Euclid's":

> Lambert appears to share the consensus view that they do not.
> It was not news, in the second half of the 18th century, that

---

[9]This is a simplification of Hilbert's more-interesting result that the biconditional [(1) iff (2)] is independent of the other axioms of order. See (Hilbert, 1899, §10).

Euclid's parallel axiom did not hold of arcs on a sphere. But
Lambert's contemporaries did not regard the arcs as lines. ...
He clearly does not take the fact that the second hypothesis is
satisfied on a spherical surface to show that it could belong to
geometry after all.[10]

In short: the sides of spherical triangles are not lines. Therefore the fact that *they* behave "non-Euclideanly" gives no reason to suppose that *lines* might.

In the next generation, we find the elegant and highly-developed non-Euclidean geometries of Lobachevsky and Bolyai. While the demonstration of rich and deep theories that deny the parallels postulate without reaching contradiction certainly provided strong support for the independence hypothesis, the existence of the theories themselves was of course no proof of that hypothesis.

In 1868, Beltrami's construction of the pseudosphere, a surface of constant negative curvature (aside from the singularity), provided a critical step in the development of views about independence.[11] Beltrami claims that his pseudosphere provides a "substrate for" Lobachevsky's geometry, and hence a demonstration that this theory is not idle. The implication immediately drawn from this observation by e.g. Hoüel is that this surface shows Lobachevsky's geometry to be consistent, which is to say that it shows the independence of the parallels postulate from the rest of Euclid.[12]

The important question for us is how, exactly, geometrical constructions like Beltrami's were taken in the late nineteenth century to show independence. Arguably, there is no single, straightforward answer to this question: the conceptual and imaginative role played by such surfaces at this point, against the background of well-developed non-Euclidean theories, was different for different thinkers. For our purposes, it will be instructive to look at one influential conception of the role of the pseudosphere and similar models, the conception shared by Helmholtz and Poincaré.

Helmholtz remarks in 1870 on the importance of Gauss's surfaces of constant non-zero curvature as follows:

> The difference between plane and spherical geometry has long
> been evident, but the meaning of the axiom of parallels could

---

[10](Dunlop, 2009, p. 47). The "second hypothesis" is the hypothesis that the internal angles of a triangle will sum to greater than two right angles.

[11](Beltrami, 1868).

[12](Hoüel, 1870). For discussion, see (Scanlan, 1988; Stump, 2007).

> not be understood till Gauss had developed the notion of surfaces flexible without dilatation and consequently that of the possibly infinite continuation of pseudospherical surfaces. Inhabiting a space of three dimensions ... we can represent to ourselves the various cases in which beings on a surface might have to develop their perception of space ...[13]

and

> These remarks will suffice to show the way in which we can infer from the known laws of our sensible perceptions the series of sensible impressions which a spherical or pseudospherical world would give us, if it existed. In doing so we nowhere meet with inconsistency or impossibility. ... We can represent to ourselves the look of a pseudospherical world in all directions ... Therefore it cannot be allowed that the axioms of our geometry depend on the native form of our perceptive faculty, or are in any way connected with it.[14]

In short, as Helmholtz sees it, these geometric constructions show that Kant is wrong. The pseudospherical model demonstrates that a non-Euclidean "world" is representable to us, with the consequence that Euclidean space is not uniquely determined by our representational capacities.

Poincaré's view in 1892 is that Beltrami has shown via his pseudosphere that no contradiction can be deduced from Lobachevsky's geometry:

> This he has done in the following manner: [I]magine beings without thickness living on [a surface of constant curvature] ... These surfaces ... are of two kinds:—Some are of *positive curvature*, and can be so deformed as to be laid on a sphere. ... Others are of negative curvature. M. Beltrami has shown that the geometry of these surfaces is none other than that of Lobachevsky.[15]

The result of this demonstration is, as Poincaré sees it, similarly anti-Kantian:

> [W]e ought ... to inquire into the nature of geometrical axioms. Are they synthetic conclusions a priori, as Kant used to

---
[13](von Helmholtz, 1876, p. 308).
[14](von Helmholtz, 1876, p. 319).
[15](Poincaré, 1892, p. 405).

say? They would appeal to us then with such force, that we could not conceive the contrary proposition, nor construct on it a theoretical edifice. There could not be a non-Euclidean geometry.[16]

That there *is* in fact a non-Euclidean geometry, as demonstrated by the combination of Lobachevsky and Beltrami, shows in Poincaré's view the conceivability of that contrary science, and hence the falsehood of the claim that Euclid's geometry is synthetic a priori.

For both Helmholtz and Poincaré, the pseudospherical surface, or perhaps that surface together with the cottage industry of producing other "non-Euclidean" surfaces, against the background of the work of Bolyai and Lobachevsky, show this: that surfaces satisfying all of Euclid aside from the parallels postulate are *possible*, and indeed *conceivable*. Similarly, presumably, for space as a whole: it is both possible and conceivable for space to be "shaped" in such a way that it satisfies all of Euclid without satisfying the parallels postulate.

It is worth noting that it is by no means a trivial inference, the inference from the existence of Beltrami-style surfaces (even in the setting of a well-developed geometry for them) to the consistency in this sense of non-Euclidean geometry. There is nothing about those surfaces that conflicts with Euclid: they are constructed within a purely Euclidean framework. By containing geodesics that satisfy most of the Euclidean postulates regarding lines, while failing to satisfy the parallels postulate, those surfaces immediately demonstrate that the parallels postulate is not semi-formally deducible from those postulates of Euclid that are satisfied by the geodesics. But unless the actual behavior of the geodesics on a pseudosphere is taken to indicate the conceivable or possible behavior of lines, this point fails to tell us anything about the conceivability or possibility of non-Euclidean surfaces or spaces. The question whether geodesics should be taken as such representatives is not a technical question but a conceptual one: Helmholtz and Poincaré, in keeping with the emerging confidence of the late 19th century in the richness and safety of the non-Euclidean framework, take it this way; those with a more conservative understanding of the nature of a line could in principle reject the inference, just as Lambert did with the sides of spherical triangles.

The acknowledgement of Beltrami's geodesics as representatives of lines underwrites the anti-Kantian conclusion of Helmholtz and of Poincaré: the

---

[16](Poincaré, 1892, pp. 406–407).

Models and Independence *circa* 1900

crucial claim is that space *could* be curved, and representable as such. From this it follows not just that the parallels postulate is not semi-formally deducible from those Euclidean postulates satisfied by the geodesics (for which conclusion we don't need the "representability" claim), but also that the parallels postulate is not provable, in a strong sense, from those postulates. For if space *could* satisfy the former without satisfying the latter, then no amount of demonstration, even appealing to the kinds of conceptual connections that loom large for Frege, will suffice to get us from one to the other.

To clarify, then: Where S is a surface constructed in Euclidean space, E the collection of Euclidean postulates satisfied by the geodesics on that surface (when we understand the term "line" in those postulates to refer not to lines but to selected geodesics on S), and Par the parallels postulate: The construction of S, together with the demonstration that its geodesics satisfy E without satisfying Par, demonstrates immediately that Par is not semi-formally deducible from E. It does not yet show that it is either possible or conceivable for *lines* to satisfy E without satisfying Par, and indeed does not show that it is possible or conceivable in any sense for lines to fail to satisfy Par, full stop. But once we add the further assumption that the geodesics on S are sufficiently line-like that their satisfaction of $(E \cup \neg Par)$ represents a real (and conceivable) possibility for lines, we obtain the stronger conclusion that a non-Euclidean space is both conceivable and possible.[17] We also, in this stronger setting, obtain the independence, in a sense strong enough to satisfy even Frege, of Par from E.

The importance of the *representative* capacity of the surface S, i.e. the representability of lines on a plane by geodesics on S, is emphasized by Poincaré in his commentary on just how such a surface undermines the claim that Euclid's axioms are synthetic a priori:

> [L]et us take a true synthetical a priori conclusion; for example, the following:—If an infinite series of positive whole numbers be taken, ..., there will always be one number that is smaller than all the others. ... Let us next try to free ourselves from this conclusion, and, denying [this proposition], to invent a false arithmetic analogous to the non-Euclidean geometry. We will find that we cannot. ...[18]

There is of course no difficulty in providing a model, in a modern sense, for the negation of the principle Poincaré describes here, the least-number

---

[17] Here, a "non-Euclidean space" is one in which lines satisfy $(E \cup \neg Par)$.
[18] (Poincaré, 1892, p. 406).

principle. Simply interpret "less-than" via the greater-than relation; alternatively, take "positive whole number" to be interpreted by the negative integers. Poincaré's claim is that we "cannot" represent *the positive whole numbers* as failing to satisfy the least-number principle; his point is that a collection that fails this principle is not the positive whole numbers. For Poincaré, as for Helmholtz, the role of a curved surface in underwriting claims about the nature of space is that it helps to demonstrate a genuine possibility for space (or rather: for planes in space) itself. No similar attempt to demonstrate the possibility of positive whole numbers that fail the least-number principle can succeed, since any collection of objects failing that principle will clearly fail to be the positive whole numbers. The difference here again is not technical but conceptual: our understanding of the nature of lines can be seen to be "loose" in a certain sense by reflection on the fact, at least as Helmholtz and Poincaré see it, that the geodesics on a curved surface can satisfy everything we demand of lines; our conception of the positive whole numbers is not sufficiently loose to allow their representation by any collection failing the least-number principle.

The kind of independence that Poincaré takes to be demonstrated by surfaces such as Beltrami's is strong: the conclusion that it is possible for lines to satisfy E without satisfying PP shows us more than just that PP is not semi-formally deducible from E; it shows additionally something about the modal properties of lines themselves. The difference is made vivid by the contrast with the least-number principle and the positive whole numbers: while it is straightforward to demonstrate the Independence$_D$, i.e. the failure of semi-formal deducibility, of that principle from, say, the denumerability of the positive whole numbers, it is, as above, not possible to show independence in Poincaré's sense. What the surface gives us in the geometric case is not (just) a means of re-interpreting the language in such a way that the given sentences express new truths and falsehoods about a new subject-matter (as is sufficient for demonstration of Independence$_D$), but in addition a pictorial representation of a situation in which lines themselves satisfy E without satisfying PP.

While the pictorial strategy proves a strong result—i.e. a result about what it is possible for lines to be like—the weakness of the strategy is that its scope of applicability is severely limited. There is no way to pictorially represent, for example, a surface on which there are counterexamples to the claim

Models and Independence *circa* 1900

(BET) If point B lies on a line between points A and C, then B lies between C and A.

And there is no straightforward way to extend the representational strategy beyond the scope of geometry.

5   The Modern era: 1899

In order for models in the style of Hilbert 1899 to be understood as proving the independence of axioms, we need to have achieved the following two developments in the conceptualization of axioms and their independence. First of all, the idea of the *independence* of a given statement from others must be divorced from the idea of the possibility, or the conceivability, of a particular arrangement of the subject-matter of those statements. Independence must instead be thought of either in terms of pure non-deducibility, or in terms of the satisfiability of abstract conditions defined by such statements when their non-logical terms are taken as place-holders. Secondly, the relation of deducibility must be understood to be "semi-formal" in the sense described above: it must be insensitive to the meanings of the geometric (or other non-logical) terms. With independence and deducibility so understood, Independence$_S$ and Independence$_D$ are straightforwardly demonstrable via Hilbert-style models.

While weaker than the kinds of independence that concerned Frege, Helmholtz, and Poincaré, the modern independence relations, Independence$_S$ and Independence$_D$, are considerably more tractable, rigorous, and broadly applicable than are the earlier notions. Models in Hilbert's sense are straightforwardly constructible for arithmetic (as in the work of Peano), and within geometry are not restricted to the independence of claims whose contrary can be visualized. A further attraction of the new method is that it goes hand in hand with the modern conception of axioms that was emerging at the turn of the nineteenth century, a conception according to which the role of axioms in mathematical theories is to characterize broad structural features of domains, rather than to express truths about determinate collections of objects and relations. As understood by e.g. Dedekind and Veblen in this period, axioms are essentially schematic, in the sense that each collection of axioms defines a complex condition satisfiable, in the typical case, by multiple domains. Given this modern understanding, the notions of independence stressed by Frege, and arguably also those stressed by Poincaré and Helmholtz, have no application to axioms: Independence$_S$

and Independence$_D$ are not only the crucial independence questions; they are the only ones that make sense.

6 The formalization of logic

In the period of roughly 1880 to 1905, it had become natural to apply to mathematical axiom-systems a handful of fundamental "meta-theoretical" questions, including those of completeness (in various senses), consistency, and mutual independence. Independence demonstrations for axioms of number theory, geometry, and analysis had become standard by the end of this period, and proceeded in essentially the Hilbert-style way.[19]

Between 1905 and 1915, the Hilbert school became increasingly interested in the development of systems of axioms for logic itself, and specifically in the formal, i.e. syntactic, specification of such systems. The natural question to raise at this point is whether the independence-proving techniques that are clearly useful in mathematical settings can be applied to systems of pure logic. Can one, for example, demonstrate the independence of logical axioms one from another using the modern technique of interpretation-style models?

Bertrand Russell, asked this question in 1909, responds as follows:

> I do not prove the independence of primitive propositions in logic by the recognized methods; this is impossible as regards principles of inference, because you can't tell what follows from supposing them false: if they are true, they must be used in deducing consequences from the hypothesis that they are false, and altogether they are too fundamental to be treated by the recognized methods.[20]

A similar sentiment appears in *Principles of Mathematics*, and again in *Principia Mathematica*.[21]

It is not difficult to see the problem as Russell sees it. Suppose (with Russell) that the basic idea of an independence proof is to assume some collection of claims to be true, assume a further claim to be false, and then check to see whether any contradiction follows. As applied to the independence of the parallels postulate, the procedure is straightforward, as is the

---

[19] See (Awodey & Reck, 2002; Zach, 1999).
[20] Russell to Jourdain, April 1909, as reported in (Grattan-Guinness, 1977, p. 117).
[21] (Russell, 1903, §17) and (Russell & Whitehead, 1910–1913, Introduction to first edition, p. 91).

result: we are assured by the coherence of the non-Euclidean surface that no contradiction follows from the supposition of $(E \cup \neg Par)$. Similarly, perhaps, for the independence of a given arithmetical axiom from others: one notes that no contradiction follows from the supposition of a handful of truths together with a target falsehood about the numbers. The role of the model in these cases is not that of re-interpretation to show Independence$_D$ or Independence$_S$; it is that of a representation of a situation in which the axioms in question are true and false respectively. Given this understanding of an independence-proof and of the role of models, it is clear that one cannot, just as Russell says, apply the technique to principles of logic. To assume a handful of logical axioms true and a target logical axiom false is already to engage in contradiction. And there is no sense in which such a collection of sentences, i.e. one including the negation of a principle of propositional logic, can be taken to describe a coherent situation. As Russell says, the principles of logic, to which we must appeal when demonstrating independence in the arenas of geometry or analysis, are "too fundamental to be treated by" such a method.

Nevertheless, there is clearly something wrong with Russell's idea that the "recognized method" is the method just described. By as early as 1905, Hilbert had already been lecturing on the use of arithmetical interpretations to show the mutual independence of axioms for propositional logic. And by 1918, Bernays uses just such interpretations to demonstrate the mutual independence of some of Russell's own propositional axioms.[22]

Bernays' technique for demonstrating the independence of an axiom A from axioms $A_1 \ldots A_n$ is essentially as follows: We give a systematic assignment of values (e.g. the numbers 1, 2, 3) to every sentence of the language, in such a way that, for example: Each of $A_1 \ldots A_n$, together with every sentence deducible from these via the specified inference rules, is assigned the value 1; A itself is not assigned 1. The immediate conclusion is that A is not deducible from $A_1 \ldots A_n$.

The first thing to note about this strategy is that, *contra* Russell, it does not involve supposing that A is false. It also doesn't involve the representation of a state of affairs that in any sense satisfies or exemplifies $A_1 \ldots A_n$; there is no need to try to make sense of the presumably-incoherent idea of a state of affairs in which an instance of an axiom of propositional logic is false. What's demonstrated by Bernays' method is simply the non-deduci-

---

[22](Bernays, 1918). For discussion, see (Zach, 1999).

bility, now in an entirely *formal*—i.e. syntactic—sense, of the formula A from the collection $A_1 \ldots A_n$ of formulas.

The difference between the method of Hilbert 1899 and Bernays 1918 can be brought out as follows. First of all, both methods are instances of what we can call the *arbitrary valuation strategy*: That strategy, applied to a formula A and a collection $A_1 \ldots A_n$ of formulas, is to assign values to formulas in such a way that, for a designated value V, V is assigned to each of $A_1 \ldots A_n$ and to each formula *deducible* from them, but is not assigned to A. The two coordinated differences between Hilbert 1899 and Bernays 1918 are (i) the kind of values employed, and (ii) the nature of the deducibility relation.

For Hilbert 1899, the value V is in each case the value "expresses a theorem of B under interpretation I," where B is a background theory (typically a theory of constructions on the real numbers) and I is the interpretation of the formulas in question via the subject-matter of B. Deducibility is a relation not explicitly specified, but understood in terms of self-evident principles of inference, subject to the constraint that the principles be semi-formal, holding independently of the interpretations of the geometric terms appearing in those formulas. The critical fact about deducibility assumed throughout is that anything deducible from a formula that expresses a theorem of B is also a formula that expresses a theorem of B. This assumed feature of deducibility is the guarantee that V is preserved by deducibility. The guarantee that the target formula A *lacks* value V, i.e. that A does not express a theorem of B under interpretation I, is given by the facts that (a) by design, $\neg A$ expresses a theorem of B under I, and (b) B is, by assumption, itself consistent.

No such valuation, and no such account of deducibility, can work in the setting of independence proofs for principles of pure logic. First of all, the designated value V must in this setting be one that is not automatically had by truths of logic (in virtue of their being truths of logic), since it is precisely a truth of logic that we will want to demonstrate lacks V. So V cannot be the value "expresses a theorem of theory B under interpretation I," for any B or I. The relation of deducibility, in addition, cannot be merely a generally-understood notion of (semi-formal) provability, since any such notion will count the formulas of pure logic as deducible from everything. Bernays' method rests on the existence of a syntactically-specified relation of deducibility, with respect to which it is not trivially true that each principle of logic is deducible from everything. It also rests on the choice of targeted value V that has nothing to do with the "interpretation," in any ordinary sense, of the formulas in question: i.e. nothing to do with the idea of those formulas as expressing truths and falsehoods about either the intended or an alternate subject-matter.

With respect to the question, then, of whether Russell is right that the "recognized method" is not applicable to questions of independence in systems of logic, the answer will turn on what exactly we take to fall under the scope of "recognized method." Taking that method to be the very broad strategy we've called the "arbitrary valuation" strategy, Russell is wrong: the method works, as Bernays shows. Taking, on the other hand, that method to be the more narrowly-construed instance of that technique in which valuations are understood in terms of re-interpretations into an assumed-consistent background theory, then Russell is right: we cannot interpret the language in such a way that axiom-sentences of propositional logic express the negations of theorems of a consistent background theory. Finally: given Russell's own, perhaps somewhat old-fashioned way of understanding the "recognized method," as involving the supposition of the falsehood of the target axiom and a subsequent question about the consistency of the result, the method is clearly not applicable to principles of logic, for the reasons Russell himself gives.

A further important difference between the method as employed by Hilbert 1899 and its refinement in Bernays 1918 involves the strength of the independence claim thereby demonstrated. As above, Hilbert's technique shows that A is not deducible from $A_1 \ldots A_n$, where "deducible from" is the semi-formal relation described above. Similarly, Bernays' technique shows that A is not deducible from $A_1 \ldots A_n$, where "deducible from" is now the rigorously-specified relation specific to a particular formal system. But Hilbert's technique, as above, also demonstrates the stronger result that we called "Independence$_S$," which is to say that it demonstrates that the condition implicitly defined by $\{A_1 \ldots A_n, \neg A\}$ is satisfiable. This it does by exhibiting a structure, namely the ordered collection of constructions out of real numbers (from the theory B), that satisfies this construction.[23] But Bernays' method provides no such further result: no domain is exhibited that satisfies conditions implicity defined by the formulas in question. And indeed, the satisfiability claim in question makes no sense as applied to the

---

[23] Strictly speaking, the existence of that structure requires not just the assumed consistency of the background theory B, as is required for the demonstration of Independence$_S$, but additionally the *truth* of B, since this is what guarantees not just the consistency, but the truth, of the claim that there exist such constructions. Hilbert's view is that, in such a setting, consistency suffices for truth. See e.g. Hilbert's letter to Frege of 29 December 1899, (Frege, 1980, p. 42).

formulas to which Bernays applies it: the axioms are not implicit definitions, and there is no sense to be made of a domain with respect to which some of those axioms express falsehoods.

The importance of Independence$_S$ is most vivid in the setting of the kind of structuralist approach to mathematical theories and axioms that was beginning to take hold at the end of the nineteenth century, and remains of central importance today. A clear early instance of this approach is found in Dedekind, for whom e.g. the theory of natural-number arithmetic is the theory of any and all ordered collections of objects that satisfy the natural-number axioms, or equivalently the theory of any and all ω-sequences.[24] The role of each axiom on this conception is to provide a partial characterization of the type of structure in question. Given a collection of axioms $A_1 \ldots A_n$, the addition of a further axiom A would be redundant if every structured domain satisfying the former collection already satisfied the latter. The important independence relation, from this point of view, is the relation of non-redundancy in this sense, which is to say that it is a matter of the satisfiability of $\{A_1 \ldots A_n, \neg A\}$, i.e. of Independence$_S$.

Independence$_D$ is the relevant kind of independence if instead the goal of the axioms is the deductive characterization of a body of truths. In this setting, A is redundant with respect to $A_1 \ldots A_n$ if A is deducible from $A_1 \ldots A_n$, and hence independent in the relevant sense if Independent$_D$ of those axioms. In a setting in which axioms are intended to provide both a deductive characterization of a theory and a definition of the structural characteristics of its domain, both kinds of independence are relevant; and as above, both are demonstrable via the construction of a model in the mode of Hilbert 1899. Once we move to Bernays 1918, however, the goal of the axioms is purely deductive (there being no sense in which the axioms of the propositional calculus define properties of structures), and the relevant kind of independence is just Independence$_D$.

7 Summing up

The idea with which we began was the traditional idea that the development of model-theoretic methods around the turn of the 20th century provided, at last, a rigorous way of answering old independence questions. The contrary proposal suggested here is that this is not quite the right way to view the developments of the period 1870–1920. Instead of a single notion of inde-

---

[24] See (Dedekind, 1888).

Models and Independence *circa* 1900

pendence that's given increasingly-rigorous treatment, we have a handful of different independence questions, some of which are susceptible to rigorous treatment, and some of which are not. As our methods have changed, so too have the questions that we are in a position to ask (and answer).

Passing from 1870 to 1899 to 1918, we see the following three lines of development.

First, we see a gradual increase in rigor. By 1899, questions of independence are divorced from questions about the representability of the subject-matter (e.g. of lines by geodesics, of positive numbers by negative numbers), and linked to the more-tractable notion of reinterpretation. By 1918, the appeal to an informal notion of provability is replaced by appeal to an explicitly-defined relation of formal deducibility.

Secondly, we see an increase in the scope of the methods. In 1870, the canonical independence-proof technique applies to the parallels postulate and to that small collection of geometric propositions whose negation can be represented as holding on something recognizably like a surface. By 1899, we have a technique that applies to all of geometry and arithmetic. And by 1918, the standard technique allows us to prove independence even of the axioms for formalized systems of logic.

But, thirdly, we see in this period a gradual decrease in the strength of the independence claims demonstrable by the emerging methods. In 1870, as understood by Helmholtz and Poincaré, a model establishes a strong modal claim, i.e. the claim that space might really be a certain way, and that we can conceive of its being that way. By 1899, the method exhibited in Hilbert's *Foundations of Geometry* makes no claim to establishing such a strong claim about the possible configuration of space or about our conception of it; the claims made via the new method are claims of non-deducibility and of the satisfiability of implicitly-defined conditions. Finally, the method employed by Bernays in 1918 provides us with clear demonstrations of non-deducibility; it is neither intended to provide, nor is it capable of providing, any modal results about the subject-matter of the theory or about the satisfiability of structural conditions.

If the oldest versions of the question of the independence of the parallels postulate are questions whose positive answer would establish the possibility or conceivability of non-Euclidean space, then they are not the questions answerable via either the method of 1899 or the method of 1918. And if the questions asked of arithmetic and geometry by Veblen and Dedekind have to do with the satisfiability of conditions implicitly defined by the axioms in question, then those questions, decisively answerable via the method of

1899, are not answerable via the method of 1918. The weakest of our independence relations, that of pure non-deducibility in a rigorously-specified formal system, is also the cleanest, and most crisply demonstrable. It is natural to take it to be, in some sense, a refinement of older more inchoate questions about independence. But if the suggestions made here are accurate, then the gap between the independence-claims cleanly demonstrable via the modern methods and the independence-claims that originally motivated much geometric work prior to the end of the 19th century is sufficiently large that we cannot take the newer methods to be merely cleaned-up ways of answering the old questions.

References

Awodey, S., & Reck, E. (2002). Completeness and Categoricity Part I: Nineteenth-century Axiomatics to Twentieth-century Metalogic. *History and Philosophy of Logic*, *23*(1), 1–30.

Beltrami, E. (1868). Saggio di Interpretazione della Geometria Non Euclidea. *Giornale di Mathematiche*, *6*, 251–288. (English translation as *Essay on the Interpretation of Non-Euclidean Geometry* by John Stillwell in (Stillwell, 1996), 7–34)

Bennett, M. K., & Birkhoff, G. (1988). Felix Klein and his 'Erlanger Programm'. In Aspray & Kitcher (Eds.), *History and Philosophy of Modern Mathematics* (pp. 145–176). Minneapolis: University of Minnesota Press.

Bernays, P. (1918). *Beiträge zur axiomatischen Behandlung des Logik-Kalküls*. Habilitation, Universität Göttingen. (Published in (Hilbert, 2013), 222–273)

Blanchette, P. (1996). Frege and Hilbert on Consistency. *The Journal of Philosophy*, *XCIII*(7), 317–336.

Blanchette, P. (2012). *Frege's Conception of Logic*. Oxford: Oxford University Press.

Dedekind, R. (1888). *Was Sind und was sollen die Zahlen*. Braunschweig. (English translation as *The Nature and Meaning of Numbers* in (Dedekind, 1901), 29–115)

Dedekind, R. (1901). *Essays on the Theory of Numbers*. Chicago: Open Court. (Edited and translated by W. W. Beman)

Dunlop, K. (2009). Why Euclid's Geometry Brooked no Doubt: J. H. Lambert on Certainty and the Existence of Models. *Synthese*, *167*,

33–65.
Frege, G. (1906). Über die Grundlagen der Geometrie. *Jahresberichte der Deutschen Mathematiker-Vereinigung, 15,* 293–309, 377–403, 423–430. (English translation as *Foundations of Geometry: Second Series* in (Frege, 1984), 293–340.)
Frege, G. (1980). *Philosophical and Mathematical Correspondence* (G. Gabriel, H. Hermes, F. Kambartel, C. Thiel, & A. Veraart, Eds.). Chicago: University of Chicago Press. (Abridged from the German edition by B. McGuinness; translated by H. Kaal.)
Frege, G. (1984). *Collected Papers on Mathematics, Logic, and Philosophy* (B. McGuinness, Ed. & M. Black et al., Trans.). Hoboken: Blackwell Publishing.
Grattan-Guinness, I. (Ed.). (1977). *Dear Russell, Dear Jourdain.* New York City: Columbia University Press.
Hilbert, D. (1899). *Grundlegung der Geometrie.* Stuttgart: Teubner. (English translation of the 10th edition: *Foundations of Geometry,* L. Unger (Trans.), P. Bernays (Ed.), Open Court 1971)
Hilbert, D. (2013). *Lectures on the Foundations of Arithmetic and Logic: 1917–1933* (W. Ewald & W. Sieg, Eds.). Berlin: Springer.
Hoüel, J. (1870). Note sur l'impossibilité de démontrer par une construction plane de la théorie des paralleles dit '*Postulatum* d'Euclide'. *Giornale di Mathematiche, 8,* 84–89.
Klein, F. (1908). *Elementarmathematik vom höheren Standpunkt aus* (Vol. 2). Berlin: Springer.
Nagel, E. (1939). The Formation of Modern Conceptions of Formal Logic in the Development of Geometry. *Osiris, 7,* 142–224.
Pesic, P. (Ed.). (2007). *Beyond Geometry Classic Papers from Riemann to Einstein.* Mineola: Dover Publications.
Poincaré, H. (1892). Non-Euclidean Geometry. *Nature, 45.*
Russell, B. (1903). *Principles of Mathematics.* Cambridge: Cambridge University Press.
Russell, B., & Whitehead, A. N. (1910–1913). *Principia Mathematica.* Cambridge: Cambridge University Press.
Scanlan, M. (1988). Beltrami's Model and the Independence of the Parallel Postulate. *History and Philosophy of Logic, 9*(1), 13–34.
Stillwell, J. (Ed.). (1996). *Sources of Hyperbolic Geometry.* Providence: American Mathematical Society.
Stump, D. (2007). The Independence of the Parallel Postulate and the Development of Rigorous Consistency Proofs. *History and Philosophy*

of Logic, *29*, 19–30.

von Helmholtz, H. (1876). On the Origin and Meaning of Geometrical Axioms. *Mind*, *3*, 302–321. (Reprinted in (Pesic, 2007), 53–70)

Zach, R. (1999). Completeness before Post: Bernays, Hilbert, and the Development of Propositional Logic. *Bulletin of Symbolic Logic*, *5*(3), 331–366.

Patricia Blanchette
University of Notre Dame
USA
E-mail: `blanchette.1@nd.edu`

Paraconsistency and Duality: between Ontological and Epistemological Views

WALTER CARNIELLI[1] AND ABILIO RODRIGUES[2]

Abstract: The aim of this paper is to show how an epistemic approach to paraconsistency may be philosophically justified based on the duality between paraconsistency and paracompleteness. The invalidity of excluded middle in intuitionistic logic may be understood as expressing that no constructive proof of a pair of propositions A and $\neg A$ is available. Analogously, in order to explain the invalidity of the principle of explosion in paraconsistent logics, it is not necessary to consider that A and $\neg A$ are *true*, but rather that there are conflicting and non-conclusive *evidence* for both.

Keywords: paraconsistent logic, philosophy of paraconsistency, evidence, intuitionistic logic

1 Introduction

Paraconsistent logics have been assuming an increasingly important place in contemporary philosophical debate. Although from the strictly technical point of view paraconsistent formal systems have reached a point of remarkable development, there are still some aspects of their philosophical significance that have not been fully explored yet. The distinctive feature of paraconsistent logics is that the principle of explosion, according to which anything follows from a contradiction, does not hold, thus allowing for the presence of contradictions without deductive triviality. Dialetheism is the view according to which there are true contradictions (cf. e.g. Berto &

---

[1]The first author acknowledges support from *FAPESP* (Fundação de Amparo à Pesquisa do Estado de São Paulo, thematic project *LogCons*) and from a *CNPq* (Conselho Nacional de Desenvolvimento Científico e Tecnológico) research grant.

[2]The second author acknowledges support from *FAPEMIG* (Fundação de Amparo à Pesquisa do Estado de Minas Gerais, research project 21308). Both authors would like to thank Henrique Almeida, Antonio Coelho, Décio Krause, André Porto, Wagner Sanz, and an anonymous referee for some valuable comments on a previous version of this text.

Priest, 2013). Paraconsistency and dialetheism are not the same thing: the latter implies the former, but one can endorse a paraconsistent logic without being dialetheist.

For those who believe in true contradictions, dialetheism does provide a philosophical justification for paraconsistency. If one accepts that reality is intrinsically contradictory, in the sense that in order to truly describe it (s)he needs some pairs of contradictory propositions, then, since reality obviously is not trivial, (s)he needs a logic in which not everything follows from a contradiction, i.e. a paraconsistent logic.

The thesis that reality is (in some sense) contradictory is an old philosophical quandary, and it has been a position defended by philosophers such as Hegel and, according to some interpreters, also by Heraclitus. This confers to contemporary dialetheism a sort of philosophical legitimacy. However, inside and outside philosophy, there is a strong reluctance in accepting that there may be entities that disobey the principle of non-contradiction as it is expressed by Aristotle in book IV of his *Metaphysics* (1005b, pp. 19–21): "the same attribute cannot at the same time belong and not belong to the same subject in the same respect" (Aristotle, 1996). Indeed, rejecting any contradiction as false is a basic methodological criterion both in sciences and in philosophy.

In order to accept contradictions without endorsing dialetheism, one needs a non-explosive negation not committed to the *truth* of a pair of propositions A and $\neg A$. Such a negation can only occur in a context of reasoning where what is at stake is a property weaker than truth, in the sense that a proposition may enjoy that property without being true. We argue in section 4 (and also in Carnielli & Rodrigues, 2016a) that the notion of *evidence*, in the sense of reasons for accepting and/or believing in a proposition, allows an epistemic reading of contradictions that is useful to describe contexts of reasoning where contradictions occur. It is perfectly feasible to imagine a scenario where non-conclusive evidence for both A and $\neg A$ is available, while no evidence for a certain B can be found, thus providing a counter-example to the principle of explosion. The notion of evidence is weaker than truth, since evidence for A does not imply that A is true, thus satisfying our requirement for a property weaker than truth to be attributed to contradictory propositions.

Truth is the central notion for classical logic. An inference is classically valid if, and only if, it preserves truth. A logic designed to represent contradictions as conflicting evidence will not be concerned with preservation of truth but, rather, with preservation of evidence. The situation is analogous to intuitionistic logic, when it is interpreted epistemically as concerned not with truth, but with the availability of a constructive proof. Our aim here is to show how such an epistemic approach to paraconsistency may be philosophically justified based on the duality between the rejection of the principle of explosion by paraconsistent and the rejection of excluded middle by paracomplete logics.

This text is structured as follows. In section 2 we discuss the duality between excluded middle and the principle of explosion. The invalidity of these inferences are the distinctive features, respectively, of paracomplete and paraconsistent logics. Section 3 shows that the rejection of the principle of excluded middle by intuitionistic logic may be understood both from an ontological and from an epistemic point of view. In section 4, an epistemic approach to paraconsistency is defended.[3]

2 On the duality between paraconsistency and paracompleteness

Let us define two n-ary logical connectives C_1 and C_2 as *dual* when

$$\sim C_1(A_1, A_2, ..., A_n) \text{ and } C_2(\sim A_1, \sim A_2, ..., \sim A_n)$$

are classically equivalent ($\sim$ meaning classical negation). Thus, $\forall$ and $\exists$, $\wedge$ and $\vee$ are dual to each other, and $\sim$ is dual to itself. We also say that the corresponding *formulas* are dual. It is clear then that the formulas

$$\sim(A \wedge \sim A) \quad \text{and} \quad A \vee \sim A$$

(expressing, respectively, non-contradiction and excluded middle) are dual. However, this duality may be seen from a different, more fundamental viewpoint, as a duality between *rules of inference*.

The principles of non-contradiction and excluded middle are often presented in books of logic, and philosophy, as fundamental laws of thought and basic tenets of classical logic. However, it is not non-contradiction but rather explosion that is essential to characterize classical negation, and this is a central feature of the classical account of logical consequence.

---

[3]This text overlaps with other papers of the authors in some points. Parts of section 2 have appeared in (Carnielli & Rodrigues, 2016b). Sections 3 and 4 develop some ideas presented in (Carnielli & Rodrigues, 2016b) and (Carnielli & Rodrigues, 2015).

Classical negation $\sim$ is defined by the following conditions (for classical $\vee$ and $\wedge$):

$$A \wedge \sim A \vDash, \tag{1}$$

$$\vDash A \vee \sim A. \tag{2}$$

Condition (1) says that there is no model M such that $A \wedge \sim A$ holds in M. (2) says that for every model M, $A \vee \sim A$ holds in M. A negation is paracomplete if it disobeys (2), and paraconsistent if it disobeys (1). Notice that each one of the conditions above corresponds exactly to half of the classical semantic clause for negation:

$$M(\sim A) = 1 \text{ if and only if } M(A) = 0. \tag{3}$$

The *only if* forbids that both A and $\neg A$ receive *1*, and the *if* forbids that both receive *0*. Given the classical definition of logical consequence (and the usual meanings of $\wedge$ and $\vee$), from (1) and (2) above it follows that for any A and B:

$$A \wedge \sim A \vDash B, \tag{4}$$

$$B \vDash A \vee \sim A. \tag{5}$$

Inference (4) above is (one version of) the principle of explosion and (5) is (again, one version of) excluded middle. Of course, excluded middle is usually presented as a valid formula or axiom, without the premise B, but this is tantamount to the formulation (5) above, which makes it clear that $A \vee \neg A$ follows from anything. It is easy to see, therefore, that from the point of view of classical logic, the fact that excluded middle is not valid in paracomplete logics and the fact that explosion is not valid in paraconsistent logics are mirror images of each other.

It is worth noting that non-contradiction and explosion are not equivalent in the following sense: one obtains a complete system of classical propositional logic by adding excluded middle and the principle of explosion to a system of posititive intuitionistic propositional logic[4], but the system so obtained turns out to be incomplete if one changes the latter to non-contradiction.

Due to the semantic clause (3) above, a central feature of classical negation is that it is a contradictory-forming operator. Applied to a proposition A, classical negation produces a proposition $\sim A$ such that A and $\sim A$ are

---

[4]Positive intuitionistic propositional logic may be defined by the usual introduction and elimination natural deduction rules for $\wedge$, $\vee$ and $\rightarrow$.

contradictories in the sense that they can neither receive simultaneously the value *0*, nor simultaneously the value *1*.

In order to give a counterexample to the principle of explosion we need a circumstance such that a pair of propositions A and $\neg A$ hold but a proposition B does not hold ($\neg$ being a paraconsistent negation). Dually, a paracomplete logic requires a circumstance such that both A and $\neg A$ do not hold (now $\neg$ is a paracomplete negation). Obviously, neither a paracomplete nor a paraconsistent negation is a contradictory-forming operator, and neither is a 'truth-functional' operator, since the semantic value of $\neg A$ is not unequivocally determined by the value of A. Now, the question is: what would be intuitive and plausible justifications for paraconsistent and paracomplete negations? An answer will be found in reasoning contexts in which negations with such characteristics occur.

3   Intuitionistic logic: a case of paracompleteness

Let us start with intuitionistic negation, which is paracomplete, as mentioned. There are two different motivations, one ontological, another epistemological, for saying that in a given context both A and $\neg A$ ($\neg$ being intuitionistic negation) do not hold.

We find in Brouwer's writings a conception of mathematical knowledge according to which there cannot be any mathematical truth not grounded on a mental construction. Furthermore, and in accordance with this thesis, the existence of a mathematical object with certain properties can be asserted only if such an object has been so constructed. Mathematics, thus, is in no way independent of thought and mind. This conception is nothing but an idealistic attitude with respect to mathematical objects: truth and existence are conceived on a idealistic basis depending on the human mind.

> *Mathematics can deal with no other matter than that which it has itself constructed.* In the preceding pages it has been shown for the fundamental parts of mathematics how they can be built up from units of perception. ... In the third chapter it will be explained why no mathematics can exist which has not been intuitively built up in this way, why consequently the only possible foundation of mathematics must be sought in this construction under the obligation carefully to watch which constructions intuition allows and which not, and why any other attempt at

> such a foundation is condemned to failure. (Brouwer, 1907, pp. 51, 73–74)

The phrase 'built up from units of perception' means that the intuition of time is the raw material from which the construction of mathematical objects proceeds. The unfolding of the process of 'two-oneness' with respect to time (or put more directly, our intuition of time, in the Kantian sense) is the base of all mathematics:

> This intuition of two-oneness, the basal intuition of mathematics, creates not only the numbers one and two, but also all finite ordinal numbers, inasmuch as one of the elements of the two-oneness may be thought of as a new two-oneness, which process may be repeated indefinitely; this gives rise still further to the smallest infinite ordinal number ω. ...
> In this way the apriority of time does not only qualify the properties of arithmetic as synthetic a priori judgements, but it does the same for those of geometry. (Brouwer, 1913, pp. 127–128)

And consequently, about mathematical truth, he says:

> [T]ruth is only in reality i.e. in the present and past experiences of consciousness ... expected experiences, and experiences attributed to others are true only as anticipations and hypotheses; in their contents there is no truth. (Brouwer, 1948, p. 488)

The rejection of excluded middle is in this setting straightforward, since it may be the case that no mental construction of A, nor of $\neg A$ has been affected. Notice, however, that according to this view, there is an identification of the notion of truth with a notion of constructive proof in the sense that a (mathematical) proposition is true if and only if a proof of it is in some sense available. If the truth of mathematical propositions is so conceived, then intuitionistic logic may still be understood as an account of truth preservation, although an idealist notion of truth. It is worth noting that this conception of mathematical truth is still compatible with a notion of truth as correspondence: what makes a mathematical proposition true is some entity that exists in reality, but in this case it is a reality constructed by thought. This is what we call an ontological motivation for rejecting excluded middle.

According to Brouwer, the classical and the constructive approaches are two irreconcilable positions in mathematics. But this is not the only way to understand intuitionistic logic. There is a weaker position, whose basic idea can already be found in Heyting, that is perfectly compatible with a realist conception of mathematical objects.

> Here, then, is an important result of the intuitionistic critique: *The idea of an existence of mathematical entities outside our minds must not enter into the proofs.* I believe that even the realists, while continuing to believe in the transcendent [transcendante] existence of mathematical entities, must recognize the importance of the question of knowing how mathematics can be built up without the use of this idea. (Heyting, 1930b, p. 306)

> [Brouwer's program] consisted in the investigation of mental mathematical construction as such, without reference to questions regarding the nature of the constructed objects, such as whether these objects exist independently of our knowledge of them. (Heyting, 1956, p. 1)[5]

Thus, an investigation of mathematical objects as mental mathematical constructions *does not need to imply* that such objects do not exist independently of such constructions, nor that they cannot be investigated by other means. In other words, one may well be a realist about mathematical objects but still have interest in intuitionistic logic as a study of such objects from the viewpoint of mental constructions.

Understood in this way, intuitionistic logic is not really about preservation of truth. Rather, it is about preservation of construction, whose specific features depend on the formal system—indeed, Heyting's and Kolmogorov's logics express different notions of construction. Given the (sup-

---

[5] It may seem strange that Heyting, a faithful disciple of Brouwer, has endorsed a position that in a certain way contradicts his master. However, the passages quoted above, and the general tone of the papers (Heyting, 1930a) and (Heyting, 1930b), make it clear that Heyting, at least in his writings, was proposing a position weaker than Brouwer's, allowing the simultaneous interest of both intuitionistic and classical logic. Perhaps Heyting's more conciliatory tone could be explained in the light of the conflict between Brouwer and Hilbert, the so-called *Grundlagenstreit*. This conflict culminated with the exclusion of Brouwer from the board of *Mathematische Annalen* in 1928. As van Dalen (1990, p. 19) puts it, "the scientific differences between the two adversaries turned into a personal animosity. The *Grundlagenstreit* is in part the collision of two strong characters, both convinced that they were under a personal obligation to save mathematics from destruction." Heyting's papers have been published in 1930, about two years after the exclusion of Brouwer. It is very plausible that Heyting, precisely because he wanted to continue to develop intuitionism, took a more careful and conciliatory position.

posed) soundness of the system, the possession of a proof of A implies the truth of A.

This position makes it possible to combine a realist notion of truth with a notion of constructive proof that is epistemic. A view that accepts a non-constructive proof of the truth of a given proposition, but distinguishes such a proof from a perhaps more informative constructive proof, is thus perfectly coherent. According to this view, what is at stake in intuitionistic logic is not an idealist notion of truth that in some way identifies proof and truth, nor some non-realist notion of truth that is constrained from an epistemic point of view. Understood in this way, the claim that in a given circumstance both A and $\neg A$ do not hold (or that both receive the semantic value 0) does not mean that both are *not true*, but only that there is no constructive proof of them, independently of the question whether any of them may be proved true by non-constructive means. The point is not the *existence* of the object, but rather the *access* to the object. Its existence may be guaranteed by a non-constructive proof, although, say, a 'direct access' may be provided only by a constructive proof.

In fact, Heyting's remarks quoted above anticipate an approach to intuitionism according to which classical and intuitionistic logics do not exclude each other. Dubucs (2008, p. 50) points out that today a "peaceful coexistence" of intuitionism and classicism has been reached and "[t]imes where controversy was raging are disappearing from collective memory". Indeed, inside mathematical departments, there is no more a dispute like that which occurred at the beginning of the twentieth century between intuitionism and formalism—the disputes are of a completely different nature. And mathematicians today, with some few exceptions, are classical and do not see any problem in applying proofs by cases based on excluded middle. However, of course it is recognized, and taken into consideration, that constructive proofs sometimes are more informative than the classical ones, since they require a *witness*: a constructive proof of an existential $\exists x F x$ demands the exhibition of an object d that satisfies Fx.

On a familiar example of a non-constructive proof

Let us illustrate this position with a familiar proposition, often mentioned as an example of a non-constructive proof:

(P) There are irrational numbers m and n such that m^n is rational.

There is a well-known non-constructive proof of (P), not acceptable to intuitionists due to an alleged illegitimate use of the excluded middle. From the supposition that

$$\sqrt{2}^{\sqrt{2}} \text{ is rational or } \sqrt{2}^{\sqrt{2}} \text{ is not rational,}$$

it can be proved that there exist numbers m and n satisfying the conditions above. Indeed, suppose that $\sqrt{2}^{\sqrt{2}}$ is rational; then $m = n = \sqrt{2}$ completes the proof. On the other hand, suppose that $\sqrt{2}^{\sqrt{2}}$ is not rational, and take $m = \sqrt{2}^{\sqrt{2}}$ and $n = \sqrt{2}$. The claim is thus proved, but we end up without knowing, after all, which are the numbers m and n that satisfy the required condition.

Now suppose that a student of mathematics, sympathetic with a realist conception of mathematical objects, becomes aware of the proof above, say, by reading section 5.1 of (van Dalen, 2008). (S)he accepts the proof as a proof of the *truth* of (P) although (s)he is still not able to exhibit the numbers m and n.

In this scenario, a non-constructive proof is available, yet a constructive proof, in this case essentially more informative, is lacking.[6]

This may be represented by the attribution of the value 0 to both P and $\neg P$, but it is important to call attention to the fact that *this does not mean that both propositions are false*. Rather, this means that neither of them has been constructively proved yet.

Our young mathematician then starts working on this problem trying to find out numbers that satisfy the required condition. After some time, (s)he constructively reaches the following result: consider $m = \sqrt{2}$ and $n = \log_{\sqrt{2}}(k)$, for k an odd natural number. It can be proved by constructive means that m^n is rational, but both m and n are irrational[7].

Fact 1 *For $k \in \mathbb{N}$, $\sqrt{2}^{\log_{\sqrt{2}}(k)}$ is a rational number.*

---

[6]Aleksandr Gelfond and Theodor Schneider independently proved in 1934 (see Baker, 1975), while solving Hilbert's 7th problem, that $2^{\sqrt{2}}$ and its square root $\sqrt{2}^{\sqrt{2}}$ are both irrational and transcendent, so in principle we know the irrational numbers m and n. Whether or not Gelfond-Schneider's proof is acceptable by intuitionists is another story.

[7]We are not claiming that m and n are constructible numbers in the intuitionistic sense. Rather, we are only exhibiting two classical real numbers that fulfil the desired properties. The intuitionistic real numbers and the classical real numbers are not comparable, and to prove that a real number *cannot be* intuitionistically defined is a hard task.

Proof. From the law of identity, $log_{\sqrt{2}}(k) = log_{\sqrt{2}}(k)$. Now, from the definition of *log*, $\sqrt{2}^{log_{\sqrt{2}}(k)} = k$, which obviously is rational, since $k \in \mathbb{N}$. □

Fact 2 $\sqrt{2}$ *is not a rational number.*

Proof. The proof in this case is well known, and it is intuitionistically acceptable, since it employs a *reductio ad absurdum* that introduces a negation. □

Fact 3 *For $k \in \mathbb{N}$ odd, $log_{\sqrt{2}}(k)$ is not a rational number.*

Proof. Suppose, recalling that by hypothesis k is odd, that $log_{\sqrt{2}}(k)$ is rational. Then, for some natural numbers a and b, $log_{\sqrt{2}}(k) = a/b$. It follows that $(\sqrt{2}^{a/b})^b = k^b$, and thus $\sqrt{2}^a = k^b$. Since k is odd, k^b is odd, for any b. Now, any natural number is odd, or not odd. Suppose a is odd. Then $\sqrt{2}^a = c\sqrt{2}$, for some c, so $\sqrt{2}^a \neq k^b$. Now, suppose a is not odd. Then, $\sqrt{2}^a = c$ for some c even. Since k^b is odd, for any b, again $\sqrt{2}^a \neq k^b$. Hence, $log_{\sqrt{2}}(k)$ is not a rational number. □

Notice that the proof above depends on the proposition *for any natural number x, x is odd or x is not odd*, which is valid intuitionistically because the predicate *x is odd* is decidable—given any natural number, there is a finite procedure that will end up with an answer *yes* or *no*. According to the clause for the universal quantifier in the *BHK* interpretation, a proof of the proposition above is a construction that transforms any given natural number d into a proof that *d is odd or d is not odd*. This is, therefore, a legitimate use of excluded middle.

Once in the possession of a constructive proof, our young mathematician says: 'now, besides knowing that the proposition is true, I am also able to exhibit numbers that make it true!' This new scenario is so represented by attributing *1* to P and *0* to $\neg P$. The rejection of the instance of the excluded middle $P \vee \neg P$, from the constructive point of view, was thus a provisional situation. It was not based on the *falsity* of both P and $\neg P$ but rather on an epistemic viewpoint related to the availability of a constructive, more informative proof. The position that accepts P as true even though it is not constructively proved is an absolutely reasonable position that illustrates the peaceful coexistence between classical and intuitionistic logic.

4 Paraconsistency and epistemological contradictions

Now, let's turn to the dual situation, the failure of explosion in paraconsistent logics. A counter-example for the principle of explosion is a circumstance in which both A and $\neg A$ hold but there is some B such that B does not hold. What would a such counter-example be? Dialetheism offers a straightforward answer: both A and $\neg A$ are true, but the world is not trivial. Although the claim that there are true contradictions does not cohere with the idea, central in scientific practice, that avoiding contradictions is an indispensable criterion of rationality, from the philosophical point of view, this answer is, in principle, defensible. The topic of 'real contradictions', or contradictions of ontological character, appears in several places in the history of philosophy. But, still, the issue is contentious. In Western philosophy, the thinkers who famously defended that reality is (in some sense) contradictory were Heraclitus and Hegel. However, for them contradictions are related to change and to movement: the ongoing motion of reality. So, it is not impossible to interpret their claims about contradictions so that the standard first-order formulation of the principle of non-contradiction (namely, the theorem-schema $\forall x \neg (Px \wedge \neg Px)$) is not violated.

However interesting a discussion on the interpretation of contradictions in Heraclitus and Hegel might be, it will not be developed here. The relevant point here is that both philosophers are usually interpreted as having given support to the thesis that there are ontological contradictions. They are taken as the philosophical background that supports contemporary dialetheism: in order to truly describe reality, we cannot dispense with some pairs of contradictory propositions. Dialetheism is what we call an ontologically motivated rejection of the principle of explosion. Nevertheless, the principle of explosion may be also rejected due to epistemological reasons.

Paraconsistency may be combined with a realist view of truth that, at the same time, endorses excluded middle and rejects the thesis that there are true contradictions. Accordingly, contradictions that occur in a number of contexts of reasoning do not mean that a given proposition A and its negation are true, nor that A is both true and false. Contradictions that occur in empirical sciences deserve special attention: it is routine for physicists to deal with theories that yield contradictions in some critical circumstances or when put together with others theories. As pointed out by Meheus (2002, p. vii), however, the fact that "almost all scientific theories at some point in their development were either internally inconsistent [i.e. contradictory] or incompatible with other accepted findings" is by no means "disastrous for good reasoning". In fact, in these cases *the general argumentative framework of science is already paraconsistent* because, obviously, in order to avoid a disaster, the principle of explosion cannot be valid. Furthermore, in such contexts it is not the case that all contradictions are equivalent. Some contradictions are, so to speak, impossible to endure, and are a sign that something has gone wrong. Others, even if understood as provisional, have to be faced and in some sense are an essential ingredient of scientific practice.

The fact that contradictions are unavoidable in empirical sciences is also pointed out by Nickles (2002, p. 2). According to him, empirical sciences are "nonmonotonic enterprises in which well justified results are routinely overturned or seriously qualified by later results. And 'nonmonotonic' implies 'temporally inconsistent' [i.e. temporally contradictory]." With respect to contradictions, he adds that they are "products of ongoing, self-corrective investigation and neither productive of general intellectual disaster nor necessarily indicative of personal or methodological failure." (Nickles, 2002, p. 2) These 'provisional contradictions' are not dialetheias. In our view, they may be of 'different kinds' in the sense of having different causes. We list some of them: (i) possible limitations of our cognitive apparatus; (ii) failure of measuring instruments and/or interactions of these instruments with phenomena; (iii) stages in the development of theories; (iv) simple mistakes that in principle could be corrected later on. In all these cases, contradictions are related primarily to knowledge and thought. This is what we call epistemic contradictions.

This idea of epistemic contradictions fits well with the conception of empirical theories as *tools to solve problems* and not correct descriptions of the world. These two approaches are discussed by Nickles (2002). The conception of theories as tools "give more attention to local problem solving and the construction of models of experiments and of phenomena than to grand unified theories" (Nickles, 2002, p. 2). Clearly, problems of consistency are, so to speak, more serious with the representational view of theories, since the latter requires that such a representation be correct (i.e. true). It is likely that some scientific theories are not only tools but may be taken as descriptions of some group of phenomena or some 'fragment' of the world, and thus can be considered representations, in the realist sense. But in contemporary science, a grand unified representation of the world is completely out of question. If this non-representational view of scientific work is accepted, and we think that it is much more plausible to understand contemporary science in this way, then contradictions that occur in the empirical sciences must be considered epistemically and not ontologically. Thus, contradictions in empirical sciences are not even potential candidates for dialetheas. They do not suggest the existence of 'contradictions in the world'. Instead, all available evidence indicates that such contradictions are epistemic.

Let us represent the fact that some contradiction is accepted in a given context by the attribution of the value *1* to a pair of propositions A and $\neg A$. Now, a question is: what does it mean to say that both A and $\neg A$ receive the value *1*, if *1* does not mean *true*? The answer must be based (as mentioned in section 1) on some property weaker than truth, in the sense that the attribution of such a property to a proposition does not imply the truth of the proposition. A pair of contradictory sentences may be understood by way of a number of concepts that are dealt with in informal reasoning: clashing information, conflicting evidence, incompatible verisimilitude, opposed possibilities, etc. Among them, the notion of *conflicting evidence* deserves special attention as particularly promising to paraconsistent logics. Evidence may be understood, as is usual in epistemology, as what is relevant for justified belief (cf. Kelly, 2014). In this sense, 'there is evidence that A is true' means that 'there are some reasons for believing that A is true' (cf. Achinstein, 2010; Kelly, 2014). Thus, evidence may be non-conclusive, and there may be evidence for the truth of A even if A is not true. Conflicting evidence occurs when one has, at the same time, reasons for accepting A and reasons for accepting $\neg A$, both non-conclusive. The notion of *preservation of evidence*, thus, presents itself as a topic to be further developed in paraconsistency. The idea is that, as much as the *BHK* interpretation for intuitionistic logic expresses preservation of (some sense of) construction, a set of inference rules and/or axioms that preserve (some sense of) evidence can be established.[8] Let us see an example of a real situation in physics that illustrates what has been said above.

The special theory of relativity

A good example of a provisional contradiction in physics is the problem faced by Einstein just before he formulated the special theory of relativity. It is well known that there was an incompatibility between the classical, Newtonian mechanics and Maxwell's theory of electromagnetic field. This is a typical case of two (supposedly) non-contradictory theories that, when put together, yield a contradiction.

---

[8] A formal system designed to express preservation of evidence can be found in Carnielli and Rodrigues (2016a).

Walter Carnielli and Abilio Rodrigues

Classical mechanics gives a description of bodies changing position in space and time. It is intuitively understood, and works very well, with respect to 'slow objects' (we will see what it means for an object to be or not to be 'slow'). Let us recall the example of a train in uniform linear motion with velocity v with respect to the rails and an object o moving inside the train with velocity w with respect to the train (where w and v have the same direction). The velocity w' of o with respect to the rails is the algebraic sum of w and v.[9] The relation between the velocities w', w and v is given by the theorem of the addition of velocities,

$$w' = w + v,$$

an elementary result in classical mechanics.

In the second half of the nineteenth century, the physicist J.C. Maxwell formulated the so-called *theory of electromagnetic field* that gives a unified account of the phenomena of electricity, magnetism and light. According to this theory, the velocity of light in vacuum (c) is equal to 300,000 km/sec, and, most importantly, c is independent of the motion of its source.

Now let us modify a bit the example above. Suppose that instead of an object moving inside the train, we are concerned with the light emitted by the headlight of the train (and suppose also, for the sake of the example, that the air has been removed). According to classical mechanics, the velocity w of the light with respect to the rails would be the sum of the velocity of the train and the velocity of light: $w = c + v$, hence, $\neg(w = c)$. On the other hand, according to Maxwell's theory, the velocity of light does not depend on the velocity of the train: $w = c$. We have, thus, that classical mechanics and the theory of electromagnetic field 'prove' a pair of contradictory propositions, $\neg(w = c)$ and $w = c$. So, the two theories put together yield a contradiction, and if the underlying logic is classical, triviality follows.

In the situation described above, two propositions A and $\neg A$ hold in the sense that both may be 'proved' from theories that were supposed to be correct. This fact may be represented by the attribution of the value *1* to both A and $\neg A$. But clearly, the meaning of this should not be that both are true—actually, we know it is not the case, and nobody has ever supposed that it could be the case. The meaning of the simultaneous attribution of the value *1*, as we suggest, is that at that time there was evidence for both in the sense, mentioned above, of some *reasons for believing* that both were

---

[9]The examples given here have been adapted from (Einstein, 1916, sections 6 and 7).

true, because there was evidence that the results yielded by both classical mechanics and the theory of electromagnetic field was true.

Classical mechanics is not compatible with Maxwell's theory because the equations of the latter are not invariant under the so-called Galilean transformations, which in classical mechanics relate the space-time coordinates of two systems of reference in uniform linear motion. By the end of the nineteenth century, H.A. Lorentz had already presented a group of equations, called Lorentz transformations, and the interesting fact is that Maxwell's equations are invariant under Lorentz transformations.[10] Einstein then rewrote Newton's equations in such a way that the theory so obtained, the theory of special relativity, was fully compatible with the theory of the electromagnetic field. Actually, what Einstein did was to consider that the mass of a body increases with velocity, and it changed the whole thing. From the new equations, a different theorem of addition of velocities can be proved:

$$w' = \frac{w+v}{1+wv/c^2}$$

The 'contradiction' is now solved (roughly speaking) in the following way: as velocity grows, time 'slows down' and 'space shortens'. So, the relation between space and time that gives velocity remains the same, because both have decreased. Newton's equations work well for 'slow objects', that is, objects moving in such a way that the value of wv/c^2 may be discarded. Thus, what the special theory of relativity shows is that classical mechanics is a special case of the former.

We have just seen a good example of what we call epistemic contradictions. We want to call attention to the fact that the general logical framework Einstein was working in was not classical. He had two different theories at hand, classical mechanics and the theory of the electromagnetic field that, when put together, yielded a non-explosive contradiction. Later, according to the special theory of relativity, the 'contradiction' disappeared. Although there was some reasons to believe that both $\neg(w = c)$ and $w = c$ were true, only one, the latter, has been established as true. The value *1* attributed to $\neg(w = c)$ later became *0*.

---

[10]We are not going into the details here. Friendly and accessible presentations of the problem may be found in (Einstein, 1916) and (a more detailed one) in (Feynman, Leighton, & Sands, 2010, ch. 15).

5 Final remarks

In the classical account there is a duality between excluded middle and explosion as rules of inference: anything follows from a contradiction, excluded middle follows from anything. Each of these rules correspond to half of the properties of classical negation. The rejection of explosion by paraconsistent logics is the mirror image of the rejection of excluded middle by paracomplete logics. We have considered here two basic motivations for paraconsistency and paracompleteness, one ontological, the other epistemological. Just as excluded middle is rejected by intuitionistic logic by reasons that may be ontological or epistemic, explosion is rejected by paraconsistent logics by reasons that may be ontological or epistemic. The ontological reasons consider that what is at stake is truth. This is the case for dialetheism, and also for intuitionistic logic from the viewpoint of Brouwer's idealism. The epistemic approach, that we find much more plausible and promising, holds that what is at stake is not truth, but the availability of evidence, in the case of paraconsistency, or the availability of a constructive proof, in the case of intuitionistic logic.

There is a sort of pragmatic argument in defense of paraconsistent logics that goes as follows. It is a fact that people reason and make inferences in contradictory contexts without trivialism. So, we need a logic able to give an account of such contexts; and so, paraconsistent logics are useful and deserve to be investigated. But a question remains: what is the nature of the contradictions handled by paraconsistent logics? We endorse the view that all contradictions are epistemic. We believe that this is the best (maybe the only) path available to provide a plausible understanding of paraconsistency.

References

Achinstein, P. (2010). Concepts of Evidence. In *Evidence, Explanation, and Realism*. Oxford University Press.

Aristotle. (1996). *The Complete Works of Aristotle*. Oxford: Oxford University Press.

Baker, A. (1975). *Transcendental Number Theory*. Cambridge: Cambridge University Press.

Berto, F., & Priest, G. (2013). Dialetheism. *Stanford Encyclopedia of Philosophy*. Retrieved from http://plato.stanford.edu/archives/sum2013/entries/dialetheism/

Brouwer, L. (1907). On the Foundations of Mathematics. In A. Heyting (Ed.), *Collected Works, Volume I*. Amsterdam: North-Holland Publishing Company.

Brouwer, L. (1913). Intuitionism and Formalism. In A. Heyting (Ed.), *Collected Works, Volume I*. Amsterdam: North-Holland Publishing Company.

Brouwer, L. (1948). Consciousness, Philosophy and Mathematics. In A. Heyting (Ed.), *Collected Works, Volume I*. Amsterdam: North-Holland Publishing Company.

Carnielli, W., & Rodrigues, A. (2015). Towards a Philosophical Understanding of the Logics of Formal Inconsistency. *Manuscrito*, *38*, 155–184.

Carnielli, W., & Rodrigues, A. (2016a). *An Epistemic Approach to Paraconsistency: a Logic of Evidence and Truth*. (Submitted paper)

Carnielli, W., & Rodrigues, A. (2016b). On the Philosophy and Mathematics of the Logics of Formal Inconsistency. In *New Directions in Paraconsistent Logic* (pp. 100–120). Berlin: Springer.

Dubucs, J. (2008). Truth and Experience of Truth. In M. van Atten et al. (Ed.), *One Hundred Years of Intuitionism*. Berlin: Birkhäuser.

Einstein, A. (1916). *Relativity: the Special and General Theory*. Emporum Books.

Feynman, R., Leighton, R., & Sands, M. (2010). *The Feynman Lectures on Physics (Volume I)*. New York City: Basic Books.

Heyting, A. (1930a). The Formal Rules of Intuitionistic Logic. In P. Mancosu (Ed.), *From Brouwer To Hilbert: the Debate on the Foundations of Mathematics in the 1920s*. Oxford: Oxford University Press.

Heyting, A. (1930b). On Intuitionistic Logic. In P. Mancosu (Ed.), *From Brouwer To Hilbert: the Debate on the Foundations of Mathematics in the 1920s*. Oxford: Oxford University Press.

Heyting, A. (1956). *Intuitionism: an Introduction*. Amsterdam: North-Holland Publishing Company.

Kelly, T. (2014). Evidence. *The Stanford Encyclopedia of Philosophy (Fall 2014)*. Retrieved from http://plato.stanford.edu/archives/fall2014/entries/evidence

Meheus, J. (2002). Preface. In J. Meheus (Ed.), *Inconsistency in Science*. Dordrecht: Springer.

Nickles, T. (2002). From Copernicus to Ptolemy: Inconsistency and Method. In J. Meheus (Ed.), *Inconsistency in Science*. Dordrecht: Springer.

van Dalen, D. (1990). The War of the Frogs and the Mice, or the Crisis of the *Mathematische Annalen*. *The Mathematical Intelligencer*, *12*(4), 17–31.

van Dalen, D. (2008). *Logic and Structure* (4th ed.). Berlin: Springer.

Walter Carnielli
CLE and Department of Philosophy—State University of Campinas
Brazil
E-mail: walter.carnielli@cle.unicamp.br

Abilio Rodrigues
Department of Philosophy—Federal University of Minas Gerais
Brazil
E-mail: abilio@ufmg.br

From Diagrammatic to Mechanical Reasoning: the Case of Syllogistic

J. MARTÍN CASTRO-MANZANO[1]

Abstract: In this contribution we sketch proofs of soundness and completeness for a diagram-based model of syllogistic that behaves as a jigsaw puzzle.

Keywords: diagrammatic logic, jigsaw puzzle, soundness, completeness

1 Introduction

In this contribution we sketch proofs of soundness and completeness for a diagram-based model of syllogistic that mechanically behaves as a jigsaw puzzle. We think this should be an interesting task because such a model has not been studied before and because, after (Allwein, Barwise, & Etchemendy, 1996) and (Shin, 1994), soundness and completeness are desiderata for *bona fide* diagrammatic logical systems. To accomplish our goal we start with a brief exposition of the notion of diagrammatic logical consequence (section 2) and then we introduce the diagrammatic system we are interested in (section 3) so that we can suggest in what sense the system is sound and complete (section 4).

2 Diagrams and diagrammatic logical consequence

Diagrams are highly estimated due to their expressive power. Notable examples of confidence in this power can be found in different historical periods (figure 1). Ramon Llull (1232–1315), for instance, is arguably the most famous case: he developed *Ars Magna*, a diagrammatic device used to *explain* divine nature to those unable to understand God's, as if diagrammatic methods were more convincing or expressive than sentential representations (figure 1a (Llull, 1501)). Thomas Murner (1475–1537) used diagrams in

---
[1] Financial support given by UPAEP Grant 30108-1008.

his *Logica Memorativa* in order to *teach* logic (figure 1b (Murner, 1509)). Dutch mathematician and philosopher of science Simon Stevin (1548–1620) developed another remarkable diagram in his *demonstration* that the efficiency of the inclined plane is a logical consequence of the impossibility of perpetual motion (figure 1c (Stevin, 1586)). And of course, besides these examples we have Descartes' (1596–1650), who is perhaps the most famous case of a philosopher (along with Wittgenstein) that made good use of diagrams in order to *model* hypothesis, such as the mechanics of the pineal gland (figure 1d (Descartes, Miller, & Miller, 1984)).

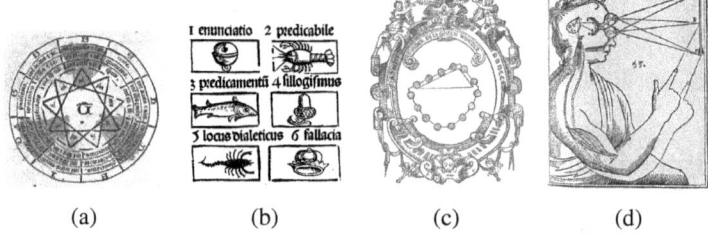

(a)   (b)   (c)   (d)

Figure 1: Notable examples of diagrams

This confidence in the power of diagrams is understandable. In order to represent knowledge we use internal and external representations. Internal representations convey mental images, for example; while external representations include physical objects on paper, blackboards, or computer screens. Following (Larkin & Simon, 1987) external representations can be divided into two classes: sentential and diagrammatic.

Sentential representations are sequences of sentences in a particular language. Diagrammatic representations are sequences of diagrams that contain information stored at one particular *locus* in a configuration, including information about relations with the adjacent *loci* (Larkin & Simon, 1987); and diagrams are information graphics[2] that index information by location

---

[2]Information graphics can be divided into quantitative charts (bar-column charts, line graphs, XY scatterplots, pie charts), maps (directional maps, topographic maps, contour maps, weather maps), tables (one way tables, two ways tables, multiway tables), pictorial illustrations, and, of course, diagrams, which can be further divided into diagrams for studying system topology (conceptual models, network diagrams), sequence and flow (flowcharts, activity diagrams), hierarchy-classification (organization charts, classification hierarchies, composition models), association (semantic networks, entity relationship diagrams), cause and effect (directed graphs, fishbone diagrams, fault tree analysis diagrams), and surely, reasoning (argument

From Diagrammatic to Mechanical Reasoning

on a plane (Larkin & Simon, 1987). The difference between diagrammatic and sentential representations, thus, is that the former preserve explicitly information about topological relations, while the latter do not—they may, of course, preserve other kinds of relations.

Hence, roughly, this confidence in the power of diagrams is comprehensible due to their computational advantages: they group together information avoiding large amounts of search, they automatically support a large number of perceptual inferences, and they grant the possibility of applying operational constraints (like *free rides* and *overdetermined alternatives* (Shimojima, 1996)) to allow the automation of perceptual inference (Larkin & Simon, 1987).

However, despite this recognized confidence, when it comes to reasoning there is a tradition (or bias?) that supports the claim that while proof-based reasoning is essential in logic and mathematics, diagram-based reasoning, no matter how useful (Nelsen, 1993) or elegant (Polster, 2004), is not, for it is not *bona fide* reasoning. Thus, for example, Tennant has suggested a diagram is only an heuristic to prompt certain trains of inference (Tennant, 1986); Dieudonné has urged a strict adherence to axiomatic methods with no appeal to geometric intuition, at least in formal proofs (Dieudonne, 2008); Lagrange remarked in the Preface to the first edition of his *Mécanique Analytique* (1788) that no figures were to be found in his work (Boissonnade, Lagrange, & Vagliente, 2013); and even Leibniz shared a similar opinion at some point (emphasis is ours):

> *La force de la démonstration est indépéndante de la figure tracée, qui n'est que pour faciliter l'intelligence de ce qu'on veut dire et fixer l'attention*; ce sont les propositions universelles, c'est-à-dire les définitions, les axiomes et les théorèmes déjà démontrés qui font le raisonnement et le soutiendraient quand la figure n'y serait pas.[3] (Bennett, Remnant, & von Leibniz, 1996, p. 309)

This claim against diagram-based reasoning is based upon the assumption that diagrams naturally lead to fallacies, mistakes, and are not susceptible of generalization. Nevertheless, we can backtrack an argument against

---

diagrams, Euler diagrams, Venn diagrams) (Nakatsu, 2009).

[3] *The cogency of demonstration is independent of the diagram, whose only role is to make it easier to understand what is meant and to fix one's attention.* It is universal propositions, i.e. definitions and axioms and theorems which have already been demonstrated, that make up the reasoning, and they would sustain it even if there were no diagram.

J. Martín Castro-Manzano

this assumption in Newton's Preface to the first edition of *Principia Mathematica* (Newton, 1979, p. 11) by reducing proof-based reasoning to mechanical reasoning (emphasis is ours):

> But as artificers do not work with perfect accuracy, it comes to pass that mechanics is so distinguished from geometry that what is perfectly accurate is called geometrical; what is less so, is called mechanical. However, *the errors are not in the art, but in the artificers*. He that works with less accuracy is an imperfect mechanic; and if any could work with perfect accuracy, he would be the most perfect mechanic of all, for the description of right lines and circles, upon which geometry is founded, belongs to mechanics.

In similar lines, Allwein et al. (1996) and Shin (1994) have developed a successful research program around heterogeneous and diagrammatic reasoning that has promoted different studies and model theoretic schemas that help us represent and better understand diagrammatic reasoning in logical terms, thus allowing a well defined study around the notions of *diagrammatic logical system* and *diagrammatic inference*.

Indeed, if reasoning is a process that produces new information given previous data and information can be represented diagrammatically and not only by way of sentences, it is not uncomfortable to suggest that diagrammatic inference is the unit of measure of diagrammatic reasoning: diagrammatic inference would be (in)correct depending on the compliance or violation of certain norms. Therefore, we could define diagrammatic logical systems with an associated notion of diagrammatic logical consequence.

For sake of explanation, let us denote the relation of a so-called diagrammatic logical consequence or diagrammatic inference between a diagrammatic configuration Δ and a singular diagram δ by $\Delta \mapsto \delta$; this relation would define our intuitions around the informal notions of *visual inference* or *visual argument* and would follow, in principle, standard structural norms: reflexivity (if $\delta \in \Delta$, $\Delta \mapsto \delta$), monotonicity (if $\Delta \mapsto \delta$, $\Delta \cup \{\delta'\} \mapsto \delta$), and cut (if $\Delta \mapsto \delta$ and $\Delta \cup \{\delta\} \mapsto \delta'$, $\Delta \mapsto \delta'$).

To give a somehow detailed account of these structural properties we could describe how the operator $\mapsto$ works by following Shimojima's definition of a *free ride*, in the short and informal version, as a process in which some reasoner gains information without following any step specifically designed to gain it (Shimojima, 1996, p. 32), in other words, a free ride is a

From Diagrammatic to Mechanical Reasoning

process that allows us to reach automatically (and sometimes inadvertently) a conclusion from a diagrammatic representation of the premises. Inversely, an *overdetermined alternative* is a situation in which a diagram that should not follow from a given diagrammatic configuration, does follow.

3 Syllogistic, jigsaw puzzles, and $L_\Box$

Before we explain the diagram-based model we are interested in, we have to explain some details about syllogistic and jigsaw puzzles in order to make the connection clear.

3.1 Syllogistic

Syllogistic is the theory of inference that deals with the consequence relation between categorical propositions. A *categorical proposition* is a proposition composed by two terms, a quantity, and a quality. The subject and the predicate of a proposition are called *terms*: the term-schema S denotes the subject term of the proposition and the term-schema P denotes the predicate. The *quantity* may be either universal (*All*) or particular (*Some*) and the *quality* may be either affirmative or negative. These categorical propositions are denoted by a *label*, either A (universal affirmative), E (universal negative), I (particular affirmative), or O (particular negative). A *categorical syllogism*, then, is a sequence of three categorical propositions ordered in such a way that two propositions are *premises* and the last one is a *conclusion*. Within the premises there is a term that appears in both premises but not in the conclusion. This particular term works as a *link* between the remaining terms and is known as the *middle term*, which we will denote with the term-schema M. According to this term we can set up four *figures* that encode and abbreviate all the valid and only the valid syllogisms (table 1).

Table 1: Valid categorical syllogisms

Figure 1	Figure 2	Figure 3	Figure 4
Barbara	*Cesare*	*Disamis*	*Calemes*
MAPSAM ⊢ SAP	PEMSAM ⊢ SEP	MIPMAS ⊢ SIP	PAMMES ⊢ SEP
Celarent	*Camestres*	*Datisi*	*Dimaris*
MEPSAM ⊢ SEP	PAMSEM ⊢ SEP	MAPMIS ⊢ SIP	PIMMAS ⊢ SIP
Darii	*Festino*	*Bocardo*	*Fresison*
MAPSIM ⊢ SIP	PEMSIM ⊢ SOP	MOPMAS ⊢ SOP	PEMMIS ⊢ SOP
Ferio	*Baroco*	*Ferison*	
MEPSIM ⊢ SOP	PAMSOM ⊢ SOP	MePMiS ⊢ SoP	

61

J. Martín Castro-Manzano

Today the connection between syllogistic and diagrams is quite clear as we can see from some notable examples in the history of logic (figure 2).

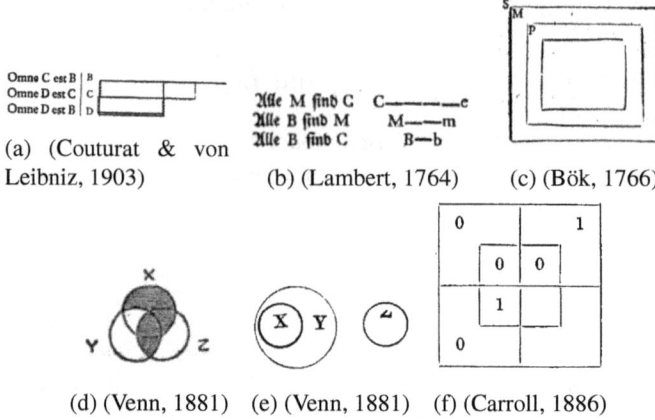

(a) (Couturat & von Leibniz, 1903) (b) (Lambert, 1764) (c) (Bök, 1766)

(d) (Venn, 1881) (e) (Venn, 1881) (f) (Carroll, 1886)

Figure 2: Notable examples of diagrams for syllogistic

3.2 Jigsaw puzzles

A *tessellation* (in the euclidean plane) is a covering of the plane without gaps or overlappings by congruent polygons. Each polygon is called a *tile* and, in particular, the simplest form of tessellation is the one that uses squares as tiles. A square-tiling, thus, is a subset of the euclidean plane that is the union of two sets of equally spaced parallel lines such that the lines of each different set are perpendicular (Meurant, 1974, p. 69). Notable and more complex examples of tessellations can be found in Islamic patterns (Critchlow, 1976), Kepler's monsters (Aiton, Duncan, Field, & Kepler, 1997), Escher's litographs (Brigham & Escher, 1978), and Penrose's aperiodic tessellations (Penrose, 1974). Now, in particular, a *jigsaw puzzle* is a tiling array composed by a finite set of tessellating pieces that require assembly by way of the *interlocking* of tiles known as *knobs* and *sockets* (figure 3): these types of puzzles date back as far as Archimedes (Botermans & Slocum, 1986, p. 13), although the typical pictorial jigsaw puzzles we are familiar with have their roots in the 1760's (Shortz & Williams, 2004, p. 4).

From Diagrammatic to Mechanical Reasoning

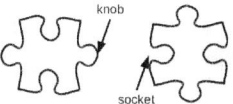

Figure 3: Pieces of a typical jigsaw puzzle

3.3 $\mathcal{L}_\square$

After this presentation of syllogistic and jigsaw puzzles consider the next analogy: just as jigsaw puzzles require the *interlocking* of tiles, syllogisms require the *linking* of terms. $\mathcal{L}_\square$ is a diagrammatic system that exploits this analogy by using a square-tiling tessellation—hence its name—in order to provide representation and a decision method for syllogistic. In this section we define $\mathcal{L}_\square$ by detailing its vocabulary, its syntax, and semantics (figure 4).

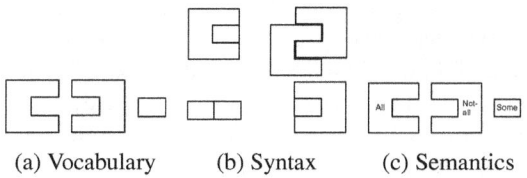

(a) Vocabulary   (b) Syntax   (c) Semantics

Figure 4: Elements of $\mathcal{L}_\square$

The vocabulary of $\mathcal{L}_\square$ is defined by two elementary diagrams (i.e. pieces or tiles), *sockets* and *knobs* (figure 4a). Syntax is given by two rules: *i)* given two elementary diagrams, the combinations of figure 4b are well formed diagrams (wfd); and *ii)* a stack of wfds is also a wfd. Semantics is given by the interpretation in figure 4c.

With these components we can represent the categorical propositions. For sake of brevity we will label each tile with an affirmative subject or predicate term-schema, S or P, assuming the tile already involves its corresponding quantity. And for sake of visualization we color the terms (figure 5a).

In the figure below we can see that this system assumes a modern interpretation of syllogistic rather than a traditional one and that the equivalence rules of *conversion*, *contraposition*, and *obversion* are all preserved by the mechanical operations of *rotating* diagrams or *switching* tiles.

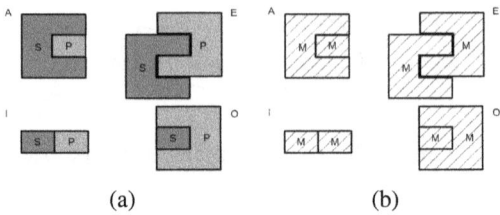

(a) (b)

Figure 5: Elements of representation of syllogistic

In order to represent categorical syllogisms and decide whether they are (in)valid we need a decision procedure, and for that we require a representation of the Identity Principle (IP). Using a single term-schema we can represent the IP in $\mathcal{L}_\square$ as follows. Consider the four categorical propositions using a single term-schema, say M, which denotes the middle term (figure 5b). It should be clear that from these four diagrammatic patterns only pattern A encodes a true statement in any structure. Then, using the IP as represented in figure 5b a linear time decision procedure for $\mathcal{L}_\square$ (Algorithm $\mathcal{A}$) takes a syllogism σ as an input and decides whether the given syllogism is (un)solvable by verifying if the concatenation of the middle terms of the given syllogism produce the IP.

Algorithm 1: Decision algorithm $\mathcal{A}$

Input: syllogism σ
if *concat(middle terms of σ)* == IP **then**
| σ is solvable;
else
| σ is unsolvable;
end

Using the previous decision procedure we can prove the 15 valid syllogisms depicted in table 1 are solvable by *stacking* up the well formed diagrams that represent the premises. When the middle term tiles *interlock* each other forming the IP (a step denoted by the arrows in figure 6), the inference is valid (i.e. there is a free ride), thus allowing the tiles S and P interlock in the third diagram, i.e. the conclusion, which is denoted below the $\mapsto$; otherwise, the inference is invalid.

From Diagrammatic to Mechanical Reasoning

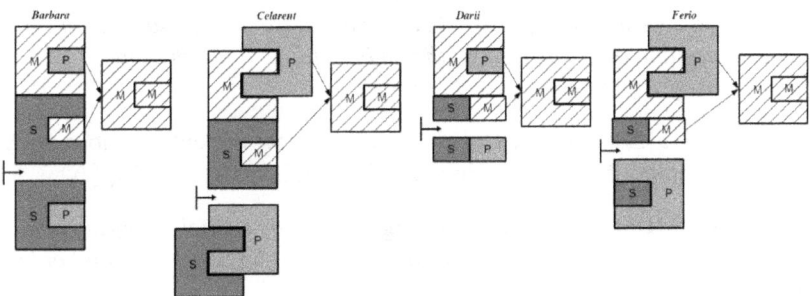

Figure 6: Proofs of syllogisms from figure 1

4 Soundness and completeness of $\mathcal{L}_\square$

Now we are in a good position to explore two fundamental logical properties of the system described above: soundness and completeness. But before we show these attributes we need some preliminary results. In first place, that:

Lemma 1 (Aristotle's Lemma for $\mathcal{L}_\square$) *Every valid syllogism is reducible to some valid syllogism from figure 1.*

From which we can infer that:

Lemma 2 (Soundness w.r.t. $\mathcal{A}$) *If $\mathcal{A}(\sigma)$ is solvable, then σ is valid.*

Lemma 3 (Completeness w.r.t. $\mathcal{A}$) *If σ is valid, then $\mathcal{A}(\sigma)$ is solvable.*

In order to state that:

Corollary 1 (Decidability) *$\mathcal{L}_\square$ is decidible.*

These results mean that $\mathcal{L}_\square$ (*jigsaw puzzle* style) allows us to obtain the right inferences (*soundness w.r.t. $\mathcal{A}$*) and only the right inferences (*completeness w.r.t. $\mathcal{A}$*) in a mechanical (*decidibility*) and efficient fashion ($O(n)$).

With these results and the following definitions we can suggest that $\mathcal{L}_\square$ is sound and complete by using a technique inspired by Shin (1994):

Definition 1 (Diagrammatic configuration) *Let h be a homomorphism from $\mathcal{L}_\square$ wfds for categorical propositions to categorical propositions in syllogistic such that $h(\delta) = \mathsf{StP}$, where δ is a wfd from $\mathcal{L}_\square$ and t is a label. A diagrammatic configuration is $Cfg(\delta) = \{h(\delta)|\delta \models h \text{ and } \delta \in \Delta\}$.*

65

Definition 2 (Single diagram) *Let δ be a wfd from $\mathcal{L}_\square$, and σ_p a categorical proposition of a syllogism σ, then δ is a single diagram representing σ_p if and only if $Cfg(\delta) = \sigma_p$.*

Thus, $Cfg(\Delta) = \bigcup_{\delta \in \Delta} Cfg(\delta)$. Hence, the set of configurations of Δ, $Cfg(\Delta)$, implies the configuration of δ, $Cfg(\delta)$, i.e. $Cfg(\delta) \subseteq Cfg(\Delta)$.

Definition 3 (Diagrammatic encoding) *Let δ be a single diagram and Δ a diagrammatic configuration, we say Δ encodes δ ($\Delta \Rrightarrow \delta$) if and only if $Cfg(\delta) \subseteq Cfg(\Delta)$.*

Definition 4 (Free ride) *Let δ be a single diagram and Δ a diagrammatic configuration, we say δ is a free ride from Δ ($\Delta \mapsto \delta$) if and only if either $\delta = \Delta$ or $\mathcal{A}(\Delta)$ is solvable (i.e. $\mathcal{A}(\Delta) = \delta$).*

We suggest that when we get a free ride in $\mathcal{L}_\square$ we encode a diagrammatic configuration, and conversely; but before we do that we require a previous result:

Lemma 4 (Validity) *Let Δ be a diagrammatic configuration of a categorical syllogism, δ_c the conclusion of Δ, and δ_i, δ_j the premises of Δ, then Δ is valid if and only if $Cfg(\delta_c) \subseteq Cfg(\delta_i) \cup Cfg(\delta_j)$.*

Proposition 1 (Soundness) *If $\Delta \mapsto \delta$, then $\Delta \Rrightarrow \delta$.*

We sketch a proof of this proposition by using induction on the length of the stack of diagrams. Let us suppose that $\Delta \mapsto \delta$. For the base case it would suffice to consider one diagram, say δ_1, which should be equal to Δ. Since $Cfg(\delta_1) = Cfg(\Delta)$, $Cfg(\delta_1) \subseteq Cfg(\Delta)$, which should imply $\Delta \Rrightarrow \delta$. For the inductive case let us suppose that if the length of the stack of diagrams that compose Δ is smaller than n, then $\Delta \Rrightarrow \delta_n$. So, there must be an application of $\mathcal{A}$ to a diagrammatic configuration Δ composed by δ_i and δ_j s.t. $i, j \leq n$. By induction hypothesis, $\Delta \Rrightarrow \delta_i$ and $\Delta \Rrightarrow \delta_j$, thus, $Cfg(\Delta)$ would encode $Cfg(\delta_i)$ and $Cfg(\delta_j)$. Now, by the soundness of $\mathcal{A}$, since $\mathcal{A}(\Delta)$ is solvable, $\mathcal{A}(\Delta) = \delta_n$, that is to say, Δ is valid, and hence $Cfg(\delta_n) \subseteq Cfg(\delta_i) \cup Cfg(\delta_j)$, which should mean that $\Delta \Rrightarrow \delta_n$.

Proposition 2 (Completeness) *If $\Delta \Rrightarrow \delta$, then $\Delta \mapsto \delta$.*

We sketch a proof of this proposition as follows. Suppose $\Delta \Rrightarrow \delta$, then $Cfg(\delta) \subseteq Cfg(\Delta)$. We may have two cases. First, $Cfg(\delta) = Cfg(\Delta)$, in which case we would have a single diagram δ s.t. $\delta = \Delta$, which would yield

$\Delta \mapsto \delta$ and is precisely what we would need. In the remaining case we would have Cfg(δ) $\subset$ Cfg(Δ). If Cfg(δ) $\subset$ Cfg(Δ) that could mean that Δ is valid. Now, if we apply $\mathcal{A}(\Delta)$, by the completeness of $\mathcal{A}$, $\mathcal{A}(\Delta) = \delta$, i.e. $\mathcal{A}(\Delta)$ is solvable, which should also imply that $\Delta \mapsto \delta$.

5 Conclusion

In this contribution we have sketched proofs soundness and completeness for a diagram-based model of syllogistic that mechanically behaves as a jigsaw puzzle. We think this is interesting because the model expounded here has not been studied before and because soundness and completeness are desiderata for any *bona fide* diagrammatic logical system.

Finally, we would like to mention that, as part of our current work, we are developing similar reconstructions for other diagrammatic logical systems (old and new, traditional and original) in order to promote the study of mechanical and diagram-based reasoning as a research program with applications, mainly, in philosophy and AI.

References

Aiton, E. J., Duncan, A. M., Field, J. V., & Kepler, J. (1997). *The Harmony of the World*. Philadelphia: American Philosophical Society.

Allwein, G., Barwise, J., & Etchemendy, J. (1996). *Logical Reasoning with Diagrams*. Oxford: Oxford University Press.

Bennett, J., Remnant, P., & von Leibniz, G. W. F. (1996). *Leibniz: New Essays on Human Understanding*. Cambridge: Cambridge University Press.

Bök, A. F. (1766). *Sammlung der Schriften, welche den logischen Calcul Herrn Ploucquets betreffen*. Frankfurt.

Boissonnade, A., Lagrange, J. L., & Vagliente, V. N. (2013). *Analytical Mechanics*. Dordrecht: Springer.

Botermans, J., & Slocum, J. (1986). *Puzzles Old and New*. Seattle: University of Washington Press.

Brigham, J. E., & Escher, M. C. (1978). *The Graphic Work of M.C. Escher*. Pan Books.

Carroll, L. (1886). *The Game of Logic*. London: Macmillan Publishers.

Couturat, L., & von Leibniz, G. W. F. (1903). *Opuscules et Fragments Inédits de Leibniz: Extraits des Manuscrits de la Bibliothèque Royale de Hanovre* (No. 1). F. Alcan.
Critchlow, K. (1976). *Islamic Patterns: an Analytical and Cosmological Approach*. New York City: Schocken Books.
Descartes, R., Miller, R. P., & Miller, V. R. (1984). *Principles of Philosophy*. Berlin: Springer.
Dieudonne, J. (2008). *Foundations of Modern Analysis*. Read Books.
Lambert, J. H. (1764). *Neues Organon*. Leipzig.
Larkin, J. H., & Simon, H. A. (1987). Why a Diagram Is (Sometimes) Worth Ten Thousand Words. *Cognitive Science*, *11*(1), 65–100.
Llull, R. (1501). *Ars Magna*. Impressum per Petru[m] Posa.
Meurant, G. (1974). *Noneuclidean Tessellations and their Groups*. Amsterdam: Elsevier.
Murner, T. (1509). *Logica Memorativa*. Argentoratum.
Nakatsu, R. T. (2009). *Diagrammatic Reasoning in AI*. Hoboken: Wiley.
Nelsen, R. B. (1993). *Proofs without Words: Exercises in Visual Thinking* (No. 1). Washington, D. C.: Mathematical Association of America.
Newton, I. (1979). *The Mathematical Principles of Natural Philosophy*. Oakland: University of California Press.
Penrose, R. (1974). The Role of Aesthetics in Pure and Applied Mathematical Research. *Bulletin of the Institute of Mathematics and its Applications*, *10*, 266ff.
Polster, B. (2004). *Q.E.D.: Beauty in Mathematical Proof*. Walker.
Shimojima, A. (1996). Logical Reasoning with Diagrams. In (chap. Operational Constraints in Diagrammatic Reasoning). Oxford: Oxford University Press.
Shin, S. J. (1994). *The Logical Status of Diagrams*. Cambridge: Cambridge University Press.
Shortz, W., & Williams, A. D. (2004). *The Jigsaw Puzzle: Piecing Together a History*. New York City: Berkeley Books.
Stevin, S. (1586). *De Beghinselen der Weeghconst*.
Tennant, N. (1986). The Withering Away of Formal Semantics? *Mind & Language*, *1*(4), 302–318.
Venn, J. (1881). *On the Diagrammatic and Mechanical Representation of Propositions and Reasonings*. R. Taylor.

From Diagrammatic to Mechanical Reasoning

J. Martín Castro-Manzano
Faculty of Philosophy and Humanities, UPAEP
Mexico
E-mail: `josemartin.castro@upaep.mx`

Game Semantics
for Vague Quantification

CHRISTIAN G. FERMÜLLER[1]

Abstract: We survey some ideas for generalizing Hintikka's semantic game for classical logic as well as Giles's game for Łukasiewicz logic to model quantification over vague predicates and with vague quantifiers. Randomized choices of admissible precisifications as well as of witnessing constants emerge as a versatile tool for modeling relevant aspects of vagueness.

Keywords: quantifiers, game semantics, vagueness, fuzzy logic

1 Introduction

How can we reason systematically with statements containing expressions like few, many, almost all, about half? Moreover, even if we stick to non-vague quantifier expressions like all and at least one: how is logical reasoning affected if the range and/or scope of the quantifier is vague? Finding adequate formal models of vague quantification is a major topic in the intersection of logic, linguistics, and the philosophy of language, with obvious links to theories of vagueness. The theory of generalized quantifiers (Peters & Westerståhl, 2006), based on classical logic, provides a suitable basis for the formal semantics of non-vague quantifiers applied in non-vague contexts. But a number of additional challenges arise for vague quantifiers as well as for non-vague quantifier expressions applied to vague predicates. Fuzzy quantifiers, based on logics over the real unit interval $[0, 1]$, have been suggested in this context; cf. (Glöckner, 2006). But usually hardly any guidance is offered on which of the myriads of *prima facie* admissible truth functions for quantifiers and other logical connectives are best suitable to model correct reasoning with vague quantifiers. Moreover many linguists, e.g. Kamp and Partee (1995), argue that (truth functional) fuzzy logic is fundamentally at odds with linguistic approaches to vagueness. To address this criticism one should come up with models that provide specific meaning to

---

[1] Supported by Austrian Science Foundation (FWF) grant I1897-N25 (MoVaQ-MFL).

degrees of truth and that allow one to derive truth functionality from first principles of reasoning, rather than to impose it without proper discussion (see, e.g., Paris, 2000).

Game semantics (Hintikka, 1968; Hintikka & Sandu, 1997) offers an analysis of logical quantifiers and propositional connectives that does not presuppose the adequateness of truth functional semantics, but rather justifies it relative to arguably more basic reasoning principles. These principle are encoded as rules of a game in which claims involving logically complex statements are systematically reduced to claims only involving atomic statements. The game can be viewed as an idealized dialogue between a proponent and an opponent, who engage in an exchange of challenges and corresponding replies, that are guided solely by the logical form of the claims at stake. The aim of this contribution is to survey some proposals that indicate that game semantics provides flexible tools for analyzing the meaning of at least some types of non-classical quantification that highlight links to fuzzy logic, but also to supervaluationist and contextualist theories of vagueness. Moreover the game approach treats quantifiers not in isolation, but rather embeds them in a semantic framework that is adequate for a wide class of logics for reasoning with graded predicates and partial truth.

2 Hintkka's game and degrees of truth

Informal dialogue games are often used to illustrate the semantics of universal and existential statements: If I claim that all objects of a domain D enjoy some property P then you are entitled to challenge me to show that d has property P for any $d \in D$ chosen by you. If, however, I state that there exists an object in D enjoying P then I may address your challenge of this claim by choosing a witness $d \in D$ with property P myself. Hintikka (1968) introduced a formal game, called $\mathcal{H}$-game here, that places this observation into the context of evaluating classical first order formulas over a given interpretation $\mathcal{I}$. The players of the $\mathcal{H}$-game are *myself* (I) and *you*, who can both act either in the role of the *proponent* **P** or of the *opponent* **O**[2] with respect to a given classical first-order formula F. For simplicity, we will assume that every (name of a) domain element is also available as constant symbol in the language. This allows us to dispense with explicit talk

---

[2]These roles of often called 'verifier' and 'falsifier', respectively, in the literature. Since we are interested in degrees of truth rather than unqualified truth here, talking about a proponent and an opponent seems more appropriate.

about variable assignments. Initially I act as **P** and you act as **O**. My aim is to show that the initial formula is true in a given interpretation $\mathcal{I}$. More generally, in any state of the game, it is **P**'s aim to show that the formula in focus at the given state, called *current formula*, is true in $\mathcal{I}$. The game proceeds according to the following rules. Note that these rules only refer to the players' roles and to the outermost connective of the current formula.

$(R_\wedge^\mathcal{H})$ If the current formula is $F \wedge G$ then **O** chooses whether the game continues with F or with G.

$(R_\vee^\mathcal{H})$ If the current formula is $F \vee G$ then **P** chooses whether the game continues with F or with G.

$(R_\neg^\mathcal{H})$ If the current formula is $\neg F$, the game continues with F, but with the players' roles switched: if I currently act as **P** and you as **O** then I will act as **O** and you as **P** at the next state, and vice versa.

$(R_\forall^\mathcal{H})$ If the current formula is $\forall x F(x)$ then **O** chooses an element d of the domain of $\mathcal{I}$ and the game continues with $F(c)$.

$(R_\exists^\mathcal{H})$ If the current formula is $\exists x F(x)$ then **P** chooses an element d of the domain of $\mathcal{I}$ and the game continues with $F(d)$.

Except for rule $R_\neg^\mathcal{H}$, the players' roles remain unchanged. The game ends when an atomic formula A is hit. The player who is currently acting as **P** *wins* and the other player, acting as **O**, *loses* if A is true in the given interpretation $\mathcal{I}$. We associate payoff 1 with winning and payoff 0 with losing. We also include the truth constants $\top$ and $\bot$, with their usual interpretation, among the atomic formulas. The game starting with formula F is called the $\mathcal{H}$-*game for* F *under* $\mathcal{I}$.

Theorem 1 (Hintikka) *I have a winning strategy in the $\mathcal{H}$-game for F under the interpretation $\mathcal{I}$ iff F is true in $\mathcal{I}$ (in symbols: $v_\mathcal{I}(F) = 1$).*

The $\mathcal{H}$-game can be played in a more general setting, where atomic formulas are not necessarily evaluated as 0 or 1, but may be assigned an arbitrary 'degree of truth' from the interval $[0, 1]$. We will speak of the $\mathcal{H}$-mv-game in that case. It matches Kleene-Zadeh logic[3] **KZ**, defined by

---

[3]Somewhat unfortunately, Kleene-Zadeh logic is sometimes simply called 'fuzzy logic'. The alternative name 'weak (fragment of) Łukasiewicz logic' will get clear in the next section.

generalizing the classical truth functions from $\{0, 1\}$ to $[0, 1]$ by extending corresponding truth value assignments to atomic formulas as follows:

$$v_\mathcal{I}(\neg F) = 1 - v_\mathcal{I}(F), \qquad v_\mathcal{I}(\bot) = 0,$$
$$v_\mathcal{I}(F \wedge G) = \min(v_\mathcal{I}(F), v_\mathcal{I}(G)), \quad v_\mathcal{I}(F \vee G) = \max(v_\mathcal{I}(F), v_\mathcal{I}(G)),$$
$$v_\mathcal{I}(\forall x F(x)) = \inf_{d \in D}(v_\mathcal{I}(F(d))), \quad v_\mathcal{I}(\exists x F(x)) = \sup_{d \in D}(v_\mathcal{I}(F(d))).$$

From now on, we will restrict attention to finite domains D only. This is quite adequate for our intended application and allows us to considerably simplify some central notions, like the following.

Definition 1 *A game has value w for player X if X has a strategy that guarantees her a payoff of at least w, while her opponent has a strategy for ensuring that X's payoff is at most w.*

It is well known from game theory that every finite, constant sum, two person game with perfect information has a unique value. In particular, we can now formulate the following.

Theorem 2 (Fermüller & Roschger, 2014) *The value for Myself of the $\mathcal{H}$-mv-game for F under the **KZ**-interpretation $\mathcal{I}$ is w iff $v_\mathcal{I}(F) = w$.*

Theorem 1 emerges as simply an instance of Theorem 2 by restricting assignments to $\{0, 1\}$ and noting that 'I have a winning strategy' expresses that the game has value 1 for me.

If one subscribes to a degree based theory of vagueness as, e.g. argued for by Smith (2008), one might be tempted to assume that logic **KZ** is a suitable basis for formalizing natural language sentences with universal and existential quantifiers involving vague predicates. But this is problematic for various reasons. In classical logic we formalize a sentence of the form All Fs are Gs as $\forall x(F(x) \to G(x))$. Since there is no connective for implication in logic **KZ**, one might think that the classically equivalent formula $\forall x(\neg F(x) \vee G(x))$ is an adequate alternative. However it is not clear at all that F or not F should be judged as logically equivalent to F implies F if F is neither fully true nor fully false. It rather seems more appropriate to argue that F implies F is always fully true, while F or not F might only be half-true if F and hence also not F are half-true. A similar worry arises for the conjunction that is implicit in a sentence like At least one child is small. While **KZ** features a conjunction connective ($\wedge$), it is not clear that it is adequate in all contexts. Hintikka's game serves to illustrate the problem: it seems natural to ask for a defense of A as well as of B in challenging the claim A and B. But the rules of the game require us to choose between a challenge of A and one of B. Fortunately, from the point of view of classical game semantics, having a winning strategy for $A \wedge B$ is equivalent to having a winning strategy for both games, one for A and one for B, if we evaluate classically. If, however, we admit intermediary degrees of truth, we should distinguish between a conjunction that requires the opponent to choose between continuing the game with either the left or the right conjunct, as in in rule $R_\wedge^\mathcal{H}$, and an alternative conjunction that entitles the opponent to cash in on the payoffs arising from subgames for both conjuncts.

3 Giles's game for Łukasiewicz logic

The $\mathcal{H}$-mv-game does not support an analysis of **All Fs are Gs** that rejects the equivocation of $F \to G$ and $\neg F \vee G$. Neither is there room for respecting the two different forms of conjunction alluded to above. To address these issues we have to consider a game similar to the $\mathcal{H}$-mv-game, except that more than just one formula may be at stake at any given state. More precisely, we will identify the state of a game with two multisets of formulas, denoted as $[F_1, \ldots, F_m \mid G_1, \ldots, G_n]$, where the multiset $\{F_1, \ldots, F_m\}$, called *your tenet*, consists of the (occurrences of) formulas currently asserted by you and the multiset $\{G_1, \ldots, G_n\}$, called *my tenet*, consist of the (occurrences of) formulas currently asserted by me. Here, 'F is asserted by player **X**' means that player **X** acts in role **P** with respect to F. A state is *final* if all formulas in it are atomic. In each non-final state one of the formula occurrences (either in my or in your tenet) is distinguished as 'current'.[4]

The following game rules are essentially due to Giles (1974):

($R_\wedge^\mathcal{G}$) If the current formula is $F \wedge G$ then the game continues in a state where the indicated occurrence of $F \wedge G$ in **P**'s tenet is replaced either by F or by G, according to **O**'s choice.

($R_\vee^\mathcal{G}$) If the current formula is $F \vee G$ then the game continues in a state where the indicated occurrence of $F \vee G$ in **P**'s tenet is replaced by either G or by F, according to **P**'s choice.

($R_\to^\mathcal{G}$) If the current formula is $F \to G$ then **O** chooses whether to simply remove it from **P**'s tenet or whether to continue the game in a state

---
[4]Strictly speaking, we only arrive at a fully defined game if we fix some 'regulation' that determines the current formula of each non-final state. Since any such regulation leads to the same value for corresponding instances of the game, we will ignore them here.

Christian G. Fermüller

where G replaces the indicated occurrence of $F \to G$ in **P**'s tenet and F is added to **O**'s tenet.

$(R_\forall^\mathcal{G})$ If the current formula is $\forall x F(x)$ then **O** chooses an element d of the domain of $\mathcal{I}$ and the game continues in a state where the indicated occurrence of $\forall x F(x)$ in **P**'s tenet is replaced by $F(d)$.

$(R_\exists^\mathcal{G})$ If the current formula is $\exists x F(x)$ then **P** chooses an element d of the domain of $\mathcal{I}$ and the game continues in a state where the indicated occurrence of $\exists x F(x)$ in **P**'s tenet is replaced by $F(d)$.

Note that, except for $R_\to^\mathcal{G}$, these rules directly generalize those of the $\mathcal{H}$-game. The implication rule $R_\to^\mathcal{G}$ encodes the following principle: if I decide to challenge your claim that F implies G then I am entitled to see you asserting G if I grant (i.e., am willing to assert) F.

As discussed at the end of Section 2, we should not only have an independent form of implication in the language, but also an alternative form of conjunction, for which we propose the following rule (Fermüller, 2009):

$(R_\&^\mathcal{G})$ If the current formula is $F \& G$ then **P** chooses whether to continue the game at a state where the indicated occurrence of $F \& G$ is replaced by both F and G in **P**'s tenet, or by a single occurrence of $\bot$, instead.

The option to assert $\bot$ instead of asserting both conjuncts is motivated by a 'principle of limited liability': the cost of asserting any single statement should never be higher than that of asserting $\bot$. Accordingly, we should always be allowed to assert $\bot$ instead of making possibly even more disadvantageous claims. While we could add this option to all other game rules uniformly, it only matters in the case where a single assertion is to be replaced by two or more assertions, assuming we play rationally.

We have not formulated a rule for negation, because we want to identify $\neg F$ with $F \to \bot$.

For final states of the game it is stipulated that the players have to pay a fixed amount of money, say 1€, for each (atomic) formula occurrence A in their tenet that is evaluated as 'false' according to an associated experiment E_A. Except for the experiments $E_\bot$ and $E_\top$ which always yield 'false' and 'true', respectively, these experiments may show dispersion: they may yield different answers upon repetition. However we assume that a fixed *risk* $\langle A \rangle$ specifies the probability that E_A results in a negative answer ('false'). Hence my overall risk, i.e. the total *expected* amount of money that I have

Game Semantics for Vague Quantification

to pay to you in the final state $[A_1, \ldots, A_n \mid B_1, \ldots, B_m]$ therefore is

$$\langle A_1, \ldots, A_n \mid B_1, \ldots, B_m \rangle = \sum_{i=1}^{m} \langle B_i \rangle - \sum_{j=1}^{n} \langle A_j \rangle.$$

The resulting game, which is called $\mathcal{G}$-game here, corresponds to Łukasiewicz logic **Ł**, which arises by augmenting the evaluation function for **KZ** with

$$v_\mathcal{I}(F \to G) = \min(1, 1 - v_\mathcal{I}(F) + v_\mathcal{I}(G)),$$
$$v_\mathcal{I}(F \& G) = \max(0, v_\mathcal{I}(F) + v_\mathcal{I}(G) - 1).$$

To be able to apply Definition 1, we stipulate that my risk at a final state of a $\mathcal{G}$-game is your payoff, while my payoff is the inverse of this risk, entailing that the game is zero-sum. Moreover we associate with each interpretation $\mathcal{I}$ a corresponding assignment of risk values defined by $\langle A \rangle_\mathcal{I} = 1 - v_\mathcal{I}(A)$ for every atomic formula A. Finally, let us speak of a $\mathcal{G}$-game for F if the initial state of the game is $[\mid F]$. These conventions allows us to formulate the characterization of Łukasiewicz logic by $\mathcal{G}$-games as follows.

Theorem 3 (Giles) *The value for Myself of a $\mathcal{G}$-game for an **Ł**-formula F under the risk value assignment $\langle \cdot \rangle_\mathcal{I}$ is $v_\mathcal{I}(F)$.*

Returning to the discussion at the end of Section 2, we note that the $\mathcal{G}$-game, in contrast to the $\mathcal{H}$-mv-game, provides a semantics for an implication that is not defined in terms of negation and disjunction and a semantics for two non-equivalent forms of conjunction. Both features are important for the formalization of binary quantifiers, as we will see below. But even if we restrict attention to unary classical quantification beyond $\forall$ and $\exists$, the more general concept of a game state in $\mathcal{G}$-games allows one to accommodate rules like the following, that do not fit the $\mathcal{H}$-game-format:

($R^{\mathcal{G}}_{\leq 2}$) If the current formula is **at most two** $x\,F(x)$ then **O** may challenge this assertion of **P** by choosing three different $d_1, d_2, d_3 \in D$, to which **P** has to reply by asserting $\neg F(a_i)$ for some $i \in \{1, 2, 3\}$.

($R_{\geq D/3}$) If the current formula is **at least a third** $x\,F(x)$ then, if challenged by **O**, **P** has to choose different $d_1, \ldots d_n \in U$ where $n \geq |D|/3$ and assert $F(d_1), \ldots, F(d_n)$.

Christian G. Fermüller

4 Addressing worries about truth functionality

Giles (1974) originally introduced his game for reasoning in theories of physics, but later Giles (1979) proposed it as a semantics for fuzzy logic. Indeed, the idea of interpreting degrees of truth as inverted risk of being proven wrong with respect to dispersive experiments that decide about the truth or falsity of atomic claims is attractive also for modeling vagueness. We might, for example, stipulate that a 'dispersive experiment' associated with the assertion Ada is tall consists in asking a randomly chosen competent speaker, who knows Ada's height, whether he or she is willing to agree with the assertion or not. Such a setup amounts to a version of voting semantics (Lawry, 1998) for fuzzy logic. Following Fermüller (2015a), we may also link the game with a contextualist approach to vagueness, where vague assertions are to be evaluated in reference to contexts that are specified by sets of admissible precisifications of the relevant atomic statements. In the simplest version of this semantics, we identify such a *context* C with a finite set of classical interpretations over the same domain and stipulate that a corresponding fuzzy interpretation, i.e. an assignment $\mathcal{I}^C$ of truth values in $[0, 1]$ to atomic formulas, is defined by[5] $v_{\mathcal{I}^C}(A) = |\{\mathcal{I} \in C : v_\mathcal{I}(A) = 1\}|/|C|$.

Dispersive experiments now appear as (uniformly) random choices of precisifications of the given context. Combined with the rules of the $\mathcal{G}$-game, this amounts to an interpretation of Łukasiewicz logic Ł that seems to justify or at least to explain truth functionality rather then to impose it directly: The $\mathcal{G}$-game rules do not directly refer to operations on truth degrees, but only to the logical structure of the current formula. Moreover, these rules seem to be justifiable independently from any particular interpretation of atomic formulas. Nevertheless a worry about truth functionality arises for (binary) quantification with vague range and scope predicates. Recall the two options for formalizing All Fs are Gs in Ł. In an interpretation $\mathcal{I}$ with domain D we obtain:

O1: $v_\mathcal{I}(\forall x(F(x) \to G(x))) = \min_{d \in D}(\min(1, 1-v_\mathcal{I}(F(d))+v_\mathcal{I}(G(d))))$,

O2: $v_\mathcal{I}(\forall x(\neg F(x) \lor G(x))) = \min_{d \in D}(\max(1 - v_\mathcal{I}(F(d)), v_\mathcal{I}(G(d))))$.

Clearly options O1 and O2 coincide for non-vague (classical) predicates. To assess their respective linguistic adequateness if applied to vague predicates consider the statement All children are nice, evaluated in a context where

---

[5]One might want to refine this by endowing the context C with a measure ν modeling the plausibility of the precisification $\mathcal{I}$ relative to the other admissible precisifications that form C.

the domain consists of persons that are all borderline cases of children and of nice persons and hence $v_\mathcal{I}(F(d)) = v_\mathcal{I}(G(d)) = 0.5$ for all $d \in D$. Following O1 the sentence is evaluated as perfectly true (1), which according to Glöckner (2006) and other fuzzy logicians is inadequate. Option O2 yields 0.5. The latter may seem more reasonable at a first glimpse; but if we evaluate All children are children instead, which yields exactly the same values in the given interpretation, then O1 appears more plausible than O2.

The indicated puzzle can be addressed by taking into account that vague range and scope predicates should not be precisified separately, but rather precisifications of the whole quantified statement should be considered.

Example 1 Let C be a context with domain $D = \{\text{Ada}, \text{Sam}\}$ that consists of the four possible precisifications that arise from independently assigning either 0 or 1 to child(d) and nice(d) for $d \in D$. Then we obtain $v_{\mathcal{I}^C}(\text{child}(d)) = v_{\mathcal{I}^C}(\text{nice}(d)) = 0.5$ for $d \in \{\text{Ada}, \text{Sam}\}$, where $\mathcal{I}^C$ is the fuzzy interpretation that corresponds to the context C, as defined above. Formalizing All children are nice and All children are children in Ł according to the two options O1 and O2, respectively, yields the following.

O1: $v_{\mathcal{I}^C}(\forall x(\text{child}(x) \to \text{nice}(x))) = v_{\mathcal{I}^C}(\forall x(\text{child}(x) \to \text{child}(x))) = 1$.

O2: $v_{\mathcal{I}^C}(\forall x(\neg\text{child}(x) \vee \text{nice}(x))) = v_{\mathcal{I}^C}(\forall x(\neg\text{child}(x) \vee \text{child}(x))) = 0.5$.

As already pointed out, both options are problematic. However, we arrive at a much more plausible result if we consider the fraction of those precisifications in C in which $\forall x(P(x) \to Q(x))$ or equivalently $\forall x(\neg P(x) \vee Q(x))$ is true: If $P = Q$ then all classical interpretations evaluate both formulas as true (1), whereas only $9/16$ of the precisifications in C evaluate the formulas as true for $P = $ child and $Q = $ nice. In other words, we obtain the truth value 1 for the formalization of All children are children and the truth value $9/16$ for the formalization of All children are nice, independently of whether we follow option O1 or O2.

The indicated style of evaluation fits our re-interpretation of dispersive experiments in Giles's game as random choices of precisifications. However the illustrated dependency of range and scope predicates calls for a mechanism that enables the players of the game to refer to precisifications not only at final game states, but already at earlier states of evaluating logically complex statements. To this aim we let each formula F that occurs at any given state possibly carry a reference to some $\mathcal{I} \in C$, denoted by $F{\uparrow}\mathcal{I}$. This enables us to specify the semantics of a new unary connective ○ by the following game rule:

Christian G. Fermüller

($R_\circ$) If the current formula is $\circ F$ then, if **O** attacks, some precisification $\mathcal{I} \in C$ is chosen randomly and $\circ F$ is replaced by $F{\uparrow}\mathcal{I}$ in **P**'s tenet.

The other game rules remain unchanged, except for stipulating that any reference $\uparrow\mathcal{I}$ is inherited from formulas to subformulas. A final state is now of the form $\Sigma = [A_1\rho_1, \ldots, A_n\rho_n \mid B_1\rho'_1, \ldots, B_m\rho'_m]$, where each ρ_i, ρ'_j is either a reference $\uparrow\mathcal{I}$ for some $\mathcal{I} \in C$ or else is empty. This yields a 'local' risk $\langle\Sigma\rangle$ by setting my risk $\langle A\rangle$ associated with any particular occurrence of A in Σ to $1 - v_\mathcal{I}(A)$ in the former case and to $\langle A\rangle_C$ in the latter case. To obtain the overall risk associated with a *non-final* state we have to take into account that all choices of precisifications are uniformly random and thus have to compute the average over corresponding local risks. Taking our clue from Theorem 3, let $\langle F\rangle_C$ denote my final overall risk with respect to context C in a game starting in state $[\mid F]$, where we both play rationally. We set $v_{\mathcal{I}^C}(F) = 1 - \langle F\rangle_C$. For the operator $\circ$, augmenting those of **Ł** we thus obtain $v_{\mathcal{I}^C}(\circ F) = |\{\mathcal{I} \in C : v_\mathcal{I}(F) = 1\}|/|C|$.

Note that every subformula in the scope of an occurrence of the connective $\circ$ is treated as a classical formula to be evaluated in the (classical) interpretations that form the given context. In particular, subformulas of the form $F \to G$ are equivalent to $\neg F \vee G$ inside the scope of $\circ$. While options O1 and O2 thus now coincide, two further possibilities for formalizing All *F*s are *G*s emerge. We will speak of 'readings' R1 and R2:

R1: $\circ\forall x(F(x) \to G(x))$ (equivalent to $\circ\forall x(\neg F(x) \vee G(x))$),

R2: $\forall x \circ (F(x) \to G(x))$ (equivalent to $\forall x \circ (\neg F(x) \vee G(x))$).

Clearly $\circ$ is redundant and the two readings coincide if F and G are classical predicates. In Example 1 we suggested that the truth value $9/16$ might be assigned to All children are nice in the specified context, since there are 9 of the 16 precisifications that form context C validate the classical formula $\circ\forall x(\mathsf{child}(x){\to}\mathsf{nice}(x))$. This amounts to formalizing the statement according to R1 as $\circ\forall x(\mathsf{child}(x){\to}\mathsf{nice}(x))$. Alternatively, we could evaluate $\mathsf{child}(\mathsf{Ada}){\to}\mathsf{nice}(\mathsf{Ada})$ and $\mathsf{child}(\mathsf{Sam}){\to}\mathsf{nice}(\mathsf{Sam})$ separately in each precisification first and find that in both cases 12 of the 16 precisifications validate the propositional formula in question. Since $\min(12/16, 12/16) = 12/16$, we obtain $12/16 = 3/4$ as truth value for All children are nice under reading R2.

We argue that, in contrast to O1 and O2, the readings R1 and R2 correspond to a genuine ambiguity that arises for universally quantified sentences with vague range and scope. A similar story arises for existentially quantified statements: while the difference between 'weak' conjunction ($\wedge$) and 'strong' conjunction ($\&$) disappears under (classical) precisifications, two different readings emerge that, in analogy to R1 and R2, can be faithfully represented formally by placing the operator ∘ either in front of or behind the existential quantifier. (A thorough discussion of this type of vagueness induced ambiguity has to be delegated to another occasion.)

5 Random witnesses and semi-fuzzy quantifiers

So far we have only considered classical (non-vague) quantifiers applied to vague predicates. To model vague proportionality quantifiers like many, about half or at least about a third we consider so-called semi-fuzzy quantification, i.e., truth functional degree based models, where the range and scope predicates are classical (bivalent). Game semantic models of semi-fuzzy quantifiers have been studied by Fermüller and Roschger (2014). Here we just outline the main ideas, focusing on the unary case; i.e. we assume that the quantifiers range over the whole domain.

In the $\mathcal{G}$-game, just like in the $\mathcal{H}$-game, universal and existential quantification is modeled by letting either the opponent **O** or the proponent **P** choose witness constants. A different type of quantifier rule arises by calling for the assertion of *random* instances of the scope formula.[6] The simplest rule for extending the $\mathcal{G}$-game in this manner specifies the basic 'random choice quantifier' Π:

($R_\Pi^\mathcal{G}$) If the current formula is $\Pi x F(x)$ then the game continues by replacing the indicated occurrence of $\Pi x F(x)$ in **P**'s tenet by a random instance of it, i.e., by $F(d)$ for some randomly chosen $d \in D$.

Remember that we are only interested in the case where the scope formula is classical and the domain D is finite, here. Under these assumptions, the following truth function for Π can be extracted from the game:

$$v_\mathcal{I}(\Pi x F(x)) = \text{Prop}_x F(x) = \frac{|\{d \in D : v_\mathcal{I}(F(d)) = 1\}|}{|D|}.$$

Since the truth value of $\Pi x F(x)$ corresponds to the proportion of domain elements that satisfy $F(x)$, Π relates to a proportional reading of many.

---

[6] All random choices are with respect to the uniform distribution over the (finite) domain, in accordance with the principle of isomorphism invariance. This is a necessary condition for the *logicality* of a quantifier, as explained by Peters and Westerståhl (2006).

Christian G. Fermüller

Suppose one is willing to accept **Many [domain elements] are** F as true if, say, at least 70% of the domain elements satisfy F. Moreover let the statement not be judged as determinately false, but rather as true to some lesser degree, if the fraction of satisfying domain elements is below 70%. Then $\overline{0.7} \to \Pi x F(x)$ is a candidate for formalizing the statement in Łukasiewicz logic enriched by Π and truth constants $\overline{x}$ where $v_\mathcal{I}(\overline{x}) = x$. In the context of the $\mathcal{G}$-game this amounts to a rule for **many** in which **O** may call on **P** to assert a random instance of the quantified statement in exchange for her (**O**'s) own assertion of an atomic statement for which the corresponding dispersive experiment is known to yield 'yes' with probability 0.7.

To assist more concise formulations of further quantifier rules we will not directly refer to the players tenets any longer, but stipulate that an *attack of F by O* entails the removal of the current formula F in **P**'s tenet. We will speak of a *bet for* F if the player asserts F, i.e. if F is added to her tenet. A *bet against* F is an assertion of $\neg F$, which amounts to asserting $\bot$ if the other player is willing to assert F. These conventions allow us to formulate rules for so-called *blind choice quantifiers*, for example for the families of quantifiers L_m^k and G_m^k:

($R_{\mathsf{G}_m^k}$) If **O** attacks $\mathsf{G}_m^k x F(x)$ then **O** bets against m random instances of $F(x)$, while **P** bets for k random instances of $F(x)$.

($R_{\mathsf{L}_m^k}$) If **O** attacks $\mathsf{L}_m^k x F(x)$ then **O** bets for k random instances of $F(x)$, while **P** bets against m random instances of $F(x)$.

Note that Π coincides with G_0^1. These rules should be considered jointly with two forms of the principle of limited liability, alluded to in Section 3.

(LLA) **O** may announce not to attack the current formula. The formula occurrence is then simply removed from **P**'s tenet (as in rule $R_\to^\mathcal{G}$).

(LLD) As a reply to any attack by **O**, **P** may choose to replace the current formula by an assertion of $\bot$ (as in rule $R_\&^\mathcal{G}$).

Truth functions that are formulated in terms of $\mathrm{Prop}_x F(x)$ can be extracted from the $\mathcal{G}$-game augmented by the rules $R_{\mathsf{G}_m^k}$ and $R_{\mathsf{L}_m^k}$, where LLA and LLD are in place.[7] The reference to $\mathrm{Prop}_x F(x)$ indicates the relation to proportionality quantifiers. For example, the willingness to bet for $2n$ random instances of $F(x)$, if the opponent agrees to bet against n such random

---

[7]The corresponding functions presented in (Fermüller & Roschger, 2014) contain an error that is corrected in (Fermüller, 2015b).

instances (for some suitable number n) might coherently be identified with a willingness to assert At least about a third [domain elements] are F. As argued in (Fermüller & Roschger, 2014), also quantifiers like about half can be modeled as blind choice quantifiers. 'Blind choice' here means that the players have to announce their bets before they get to know which domain elements have been (randomly) chosen. An interesting alternative for formulating semantic game rules for quantifiers arises if the identity of randomly picked domain elements is revealed to the players *before* they place corresponding bets. A simple example of such a rule for a *deliberate choice quantifier* is:

($R_{\Pi_1^1}$) If **O** attacks $\Pi_1^1 x F(x)$ then 2 domain elements d_1 and d_2 are chosen randomly and **P** may choose to bet either for $F(d_1)$ and against $F(d_2)$ or against $F(d_1)$ and for $F(d_2)$.

This rule matches $v_\mathcal{I}(\Pi_1^1 F(x)) = 2(\text{Prop}_x F(x))(1 - \text{Prop}_x F(x))$.

The common feature of all such models is the interpretation of the degree of truth of a quantified statement as the inverse of the overall risk induced by bets either for or against random instances of the statement, possibly in combination with further risk increasing or risk decreasing operators. We refer to (Fermüller & Roschger, 2014) for a systematic analysis of such quantifiers, including their suitability for modeling natural language expressions like about half or about three in ten.

6 Fully vague quantification

In Section 4 we have seen that adequate models of classical (non-vague) quantifiers, like all, applied to vague predicates call for the evaluation with respect to contexts of precisifications in a manner that reflects possible dependencies between range and scope predicates. This was achieved by introducing the operator ∘, triggering a random choice of precisification. In Section 5 we modeled vague quantifiers like about half applied to classical predicates via rules that involve random choices of witness constants. Clearly these two types of randomization are related and should be combined when vague quantifiers are applied to vague predicates.

The game rules for semi-fuzzy quantifiers reviewed in Section 5 could, in principle, be applied to vague (graded) predicates as well, since the rules just reduce the evaluation of the current formula to that of subformulas. But that would imply that $\text{Prop}_x F(x)$ cannot be any longer understood as the proportion of domain elements satisfying F and hence it is not clear whether the resulting truth functions match pre-formal intuitions about correct reasoning involving fully vague quantification. Arguably more adequate models arise by applying the quantifier rules only to admissible precisifications of scope formulas. This can be achieved by simply placing the ○-connective in front of the corresponding formula. But note that, in contrast to the case for ∀ discussed in Section 4, an evaluation in a precisification (i.e. a classical interpretation of the predicates) will return an intermediate truth value in general. Therefore the operator ○ now adopts a more general meaning: it does not return the fraction of interpretations making the formula (fully) true, but rather the *average* of—possibly intermediate—truth values arising in admissible precisifications.

For semi-fuzzy quantifiers we made the simplifying assumption that the range of quantification coincides with the domain. In contrast to universal and existential quantification, we cannot, in general, use propositional connectives to formalize the relativization of the domain to elements satisfying the range predicate. However, as explained in (Fermüller & Roschger, 2014) such a relativization can uniformly be modeled game semantically by employing subgames for determining whether a chosen element is in the range of quantification, if the range predicates are classical. Vague range predicates again call for preceding precisification as modeled by ○.

In any case, we argue that, in contrast to traditional models based on fuzzy logic, game semantics supports the conceptual separation of two different sources of uncertainty involved in vague quantification: (1) uncertainty about the precise meaning of quantifiers expressions and (2) uncertainty relating to the vagueness of involved predicates.

7 Conclusion

We hope to have shown that Giles-style games provide a powerful and versatile tool for modeling the semantics of vague quantification. We conclude by pointing out a feature of our approach, that has not yet been discussed explicitly: game based semantics involving random choices of precisifying interpretations as well as of witnessing constants exhibits considerable robustness with respect to various potential deficiencies of the players' knowledge. The semantic games can be meaningfully played even if neither the set of admissible precisifications nor the domain is completely known to the players. We submit that this feature is very welcome, if not essential in aiming for linguistically adequate models of logical reasoning with sentences like Many people are afraid of refugees and Very few strangers seek to harm their hosts in realistic contexts.

References

Fermüller, C. G. (2009). Revisiting Giles's Game. In *Games: Unifying Logic, Language, and Philosophy* (pp. 209–227). Berlin: Springer.

Fermüller, C. G. (2015a). Combining Fuzziness and Context Sensitivity in Game Based Models of Vague Quantification. In *Integrated Uncertainty in Knowledge Modelling 2015, Lecture Notes in Computer Science 9376* (pp. 19–31). Berlin: Springer.

Fermüller, C. G. (2015b). Semantic Games for Fuzzy Logics. In P. Cintula, C. Fermüller, & C. Noguera (Eds.), *Handbook of Mathematical Fuzzy Logic, Volume 3* (pp. 969–1028). London: College Publications.

Fermüller, C. G., & Roschger, C. (2014). Randomized Game Semantics for Semi-fuzzy Quantifiers. *Logic Journal of the Interest Group of Pure and Applied Logic, 22*(3), 413–439.

Giles, R. (1974). A Non-classical Logic for Physics. *Studia Logica, 33*(4), 399–417.

Giles, R. (1979). A Formal System for Fuzzy Reasoning. *Fuzzy Sets and Systems, 2,* 233–257.

Glöckner, I. (2006). *Fuzzy Quantifiers: a Computational Theory*. Berlin: Springer.

Hintikka, J. (1968). Language-games for Quantifiers. In N. Rescher (Ed.), *Studies in Logical Theory* (pp. 46–72). Hoboken: Blackwell Publishing.

Hintikka, J., & Sandu, G. (1997). Game-theoretical Semantics. In A. ter Meulen & J. van Benthem (Eds.), *Handbook of Logic and Language* (pp. 361–410). Amsterdam and Cambridge, Massachusetts: Elsevier and the MIT Press.

Kamp, H., & Partee, B. (1995). Prototype Theory and Compositionality. *Cognition, 57,* 129–191.

Lawry, J. (1998). A Voting Mechanism for Fuzzy Logic. *International Journal of Approximate Reasoning, 19*(3–4), 315–333.

Paris, J. (2000). Semantics for Fuzzy Logic Supporting Truth Functionality. In V. Novák & I. Perfilieva (Eds.), *Discovering the World with Fuzzy Logic* (pp. 82–104). Berlin: Springer.

Christian G. Fermüller

Peters, S., & Westerståhl, D. (2006). *Quantifiers in Language and Logic.* Oxford: Clarendon Press.

Smith, N. J. (2008). *Vagueness and Degrees of Truth.* Oxford: Oxford University Press.

Christian G. Fermüller
Vienna University of Technology
Austria
E-mail: `chrisf@logic.at`

A Nonmonotonic Sequent Calculus for Inferentialist Expressivists

ULF HLOBIL[1]

Abstract: I am presenting a sequent calculus that extends a nonmonotonic consequence relation over an atomic language to a logically complex language. The system is in line with two guiding philosophical ideas: (i) logical inferentialism and (ii) logical expressivism. The extension defined by the sequent rules is conservative. The conditional tracks the consequence relation and negation tracks incoherence. Besides the ordinary propositional connectives, the sequent calculus introduces a new kind of modal operator that marks implications that hold monotonically. Transitivity fails, but for good reasons. Intuitionism and classical logic can easily be recovered from the system.

Keywords: nonmonotonic logic, sequent calculus, logical inferentialism, logical expressivism, material consequence relation

1 Philosophical motivation

What follows is motivated by two big philosophical ideas: logical inferentialism and logical expressivism. Logical inferentialism is a view about the meaning of logical vocabulary. Very roughly, it says that the meaning of logical vocabulary is settled by its inferential role, i.e., by what implies and is implied by sentences in which such vocabulary occurs. Logical expressivism is a view about the expressive function of logical vocabulary, i.e., a view about what such vocabulary is for, what it allows us to do. Very roughly, the view is that logical vocabulary allows us to explicitly undertake commitments regarding inferential goodness and incoherence by asserting logically complex sentences, whereas without logical vocabulary we could undertake such commitments only implicitly by reasoning or arguing in certain ways. It is part of this idea that we can introduce logical vocabulary

---

[1]The work I am presenting here comes out of joint work with Robert Brandom and his research group on nonmonotonic logic. So my debt to Robert Brandom and the other members of the group can hardly be overestimated. I also want to thank the participants of the Logica 2015 conference for invaluable comments and discussion.

purely in terms of a material consequence relation and incoherence property over a language that does not include logical vocabulary. I shall present a logical system that exemplifies logical inferentialism and logical expressivism. The system introduces logical vocabulary in terms of its inferential role, and it does so on the basis of material consequence and incoherence. The perhaps biggest challenge for such a project is that material, nonlogical consequence and incoherence are virtually always nonmonotonic. Nonmonotonicity, however, is notoriously difficult to deal with in formal systems. In this section, I want to explain the basic ideas just mentioned.

1.1 Logical inferentialism

Let's begin with logical inferentialism. This is the view that the meaning of logical vocabulary is a matter of its inferential role (for a recent exposition and defense see Peregrin, 2014). Gentzen (1934, p. 189) formulated a version of the idea when he famously said that the introduction rules of a bit of logical vocabulary constitute, "as it were, the 'definitions' of the symbols concerned." The version of the idea that will be relevant here, however, is closer to Dummett's (1991, p. 247) view that the "meaning of [a] logical constant [...] can be completely determined by laying down the fundamental logical laws governing it" (see also Kneale, 1956, pp. 254–55). For our current purposes, we can think of logical inferentialism as the idea that the meaning of a bit of logical vocabulary is fully determined by the full set of implications or good arguments in which it occurs. Hence, we can introduce such vocabulary into a language by giving rules that determine the consequence relation over the logically extended language. Below I will provide such rules in the form of a sequent calculus.

Logical inferentialism has been criticized in various ways. Entering such debates here would take us too far afield. The only point that will matter for me is so-called "conservativeness." Prior (1960) famously pointed out that one can introduce connectives, like his "tonk," that trivialize a consequence relation by laying down introduction and elimination rules. Supposing that such connectives are meaningless, this can seem to undermine inferentialism because it shows that not all rules that specify an inferential role specify a meaning. In response to this worry, most inferentialists follow Belnap (1962) and say that the rules by which we introduce a new bit of logical vocabulary must extend the consequence relation we start with in a conservative manner. That is, an implication that does not contain the new bit of vocabulary holds in the extended consequence relation just in case it already held in the unextended consequence relation. I accept this as a restriction on the rules we can use to introduce logical vocabulary. Many more such restrictions have been proposed in the literature, such as various versions of harmony and separability. However, I will ignore such further restrictions here and shall be content with the safeguard that conservativeness provides against 'tonk-like' connectives.

1.2 Logical expressivism

I am taking the idea of logical expressivism from Robert Brandom (together with whom I have developed the ideas presented here). Brandom builds on Frege's idea that his "concept-script is a formal language for the explicit codification of conceptual contents" (Brandom, 2000, p. 58). If one is (with Brandom) an inferentialist across the board and not just regarding logical vocabulary, one believes that all (non-logical) conceptual contents are a matter of material consequence and incoherence. On this view, Frege's idea is that the concept-script is a formal language for the explicit codification of material consequence and incoherence. Hence, the expressive function of a formal language is to let us talk 'about' material implication relations and incoherence properties.[2] Brandom sometimes puts this view in a slogan by saying that logic "is the organ of semantic self-consciousness" (Brandom, 2009, p. 11).

Logical expressivism would need a lot of unpacking, but for our purposes, we can simplify the idea to the claim that, for any well-behaved language, logical vocabulary can be introduced solely in terms of the material consequence relation and incoherence property of the unextended language, and the so introduced logical vocabulary allows us to make explicit this consequence relation and incoherence property within the object language.

Definition 1 *Logical expressivism is the thesis that (i) logical vocabulary can be introduced into any language with a well-behaved material consequence relation and incoherence property solely in terms of this consequence relation and incoherence property, and (ii) the thus introduced vocabulary allows us to form sentences that make explicit facts about the underlying (and also the extended) consequence relation and incoherence property.*

---

[2]Notice that, given logical inferentialism, the "about" here must not (or not primarily) be understood in representationalist terms.

For us, the first point concerns the raw materials that we start with: a material consequence relation and incoherence property defined over a language without logical vocabulary.

The second point is more difficult to understand. It concerns what we want to built from the basic material: we want to build logical expressions that fulfill their expressive job of making explicit consequence and incoherence. Now, when can a bit of vocabulary count as "making explicit" the material consequence relation and incoherence property? This is easiest to grasp for the two logical expressions that I take to be paradigmatic: the conditional and negation. In the system I will present below, the conditional makes explicit—or tracks—consequence, and the negation makes explicit—or tracks—incoherence. For the conditional, this means that a conditional $A \to B$ is implied by a premise-set just in case B is implied by the result of adding A to this premise-set, i.e., a deduction theorem holds. For negation, it means that the negation $\neg A$ is implied by a premise-set just in case adding A to this premise-set results in something incoherent. So logical expressivism puts constraints on the conditional and negation that are acceptable for us.

1.3 Nonmonotonicity

Brandom and Aker have already provided a system in which logical vocabulary is introduced solely in terms of a material incoherence property over sets of atomic sentences (Brandom, 2008, pp. 141–175). One crucial limitation of this so-called "incompatibility semantics" is that it's consequence relation is monotonic, i.e., if a set of sentences implies, say, the sentence A then so do all its supersets (see lemma 2.1 on p. 143 of Brandom, 2008).

This is a limitation because, according to inferentialist expressivism regarding logic, material inferences are not just enthymematic formal inferences. If we take such inferences at face value, however, it is hard to see how their nonmonotonicity could be merely apparent. Moreover, paradigmatic material implications, such as implications in legal matters, medicine, or morality, are virtually always defeasible. And the same holds for material incoherence. Sets of commitments that don't fit together can become jointly acceptable once we add another commitment into the mix.

Is there perhaps another off-the-shelf logic that suits the inferentialist expressivists as an exemplification of her ideas? Unfortunately, it does not seem so. There are many nonmonotonic logics on offer today (for an introductory overview see Antonelli, 2008). But, as far as I know, none of them uses a material consequence relation and incoherence property as their

A Nonmonotonic Sequent Calculus

starting point. In fact, many nonmonotonic logics treat some logic—often classical logic—and its vocabulary as given and freely available in the new logic. Moreover, most nonmonotonic logics obey some version of Cut. As we will see below, this means that they cannot have a conditional that is in line with logical expressivism, i.e., they cannot have a deduction theorem.

If logical inferentialism and logical expressivism are good ideas and we take the nonmonotonicity of material consequences seriously, there should be formal systems that exemplify these ideas in a paradigmatic way. Thus, we want a way of conservatively extending a nonmonotonic material consequence relation and incoherence properties such that the conditional and negation track consequence and incoherence, respectively.

2 The basic setup

As I explained in the previous section, our motivating philosophical ideas are, firstly, that the meaning of logical vocabulary is determined by its inferential role and, secondly, that logical vocabulary makes explicit features of an underlying, nonmonotonic, material consequence relation and incoherence property. So we must start with a material consequence relation and incoherence property over a language that does not contain logical vocabulary. Call this language $\mathscr{L}_{0-}$. We can think of $\mathscr{L}_{0-}$ as a set of atomic sentences, $\{p_1, \ldots, p_n\}$. Some subsets of $\mathscr{L}_{0-}$ materially imply some sentences in $\mathscr{L}_{0-}$. And some subsets of $\mathscr{L}_{0-}$ are materially incoherent. So the structures that we start with are triples of (a) an atomic language, (b) a (single conclusion) consequence relation over it, and (c) an incoherence property defined over sets of atoms.

In order to express incoherence and consequence in a unified way, we introduce the constant "$\bot$." Let $\mathscr{L}_0 = \mathscr{L}_{0-} \cup \{\bot\}$. Let $\mathrel{|\!\sim}_0$ be the relation over $\mathscr{L}_0$ such that $\{p_k, \ldots, p_l\} \mathrel{|\!\sim}_0 p_i$ iff $\{p_k, \ldots, p_l\}$ materially implies p_i and, moreover, such that $\Gamma \mathrel{|\!\sim}_0 \bot$ iff Γ is materially incoherent.[3] The constant $\bot$ cannot occur on the left of the snake-turnstile and it cannot be embedded. We used it merely to encode the incoherence property into the "underlying consequence relation;" so $\mathrel{|\!\sim}_0 \subseteq \mathcal{P}(\mathscr{L}_{0-}) \times \mathscr{L}_0$. We say that a consequence relation $\mathrel{|\!\sim}_0$ is proper just in case (a) the whole atomic language is incoherent ($\mathscr{L}_{0-} \mathrel{|\!\sim}_0 \bot$), (b) the empty set is coherent ($\emptyset \mathrel{|\!\not\sim}_0 \bot$), (c) $\mathrel{|\!\sim}_0$ is

---

[3] I use capital Greek letters for sets of sentences, lower case Latin letters for atomic sentences, and upper case Latin letters for arbitrary sentences. I will omit the set-brackets on the left of the snake-turnstile if no confusion can arise.

reflexive ($\forall \Gamma \subseteq \mathscr{L}_{0-} (p \in \Gamma \Rightarrow \Gamma \mathrel{\vert\!\sim}_0 p)$), and (d) $\mathrel{\vert\!\sim}_0$ obeys what we call "Ex Falso Fixo Quodlibet" (ExFF):

ExFF   For any atom p, if $\forall \Delta \subseteq \mathscr{L}_{0-} (\Gamma, \Delta \mathrel{\vert\!\sim}_0 \bot)$, then $\Gamma \mathrel{\vert\!\sim}_0 p$.

This principle is a variant of *ex falso quodlibet*, i.e., explosion. Notice that the difference between the traditional version of *ex falso* and ExFF only shows up in a nonmonotonic context. After all, $\Gamma \mathrel{\vert\!\sim}_0 \bot$ guarantees $\forall \Delta \subseteq \mathscr{L}_{0-} (\Gamma, \Delta \mathrel{\vert\!\sim}_0 \bot)$ if monotonicity holds for $\mathrel{\vert\!\sim}_0$.

Let's sum up our starting point in two definitions.

Definition 2 *Base Structure: A base structure is a pair $\langle \mathscr{L}_0, \mathrel{\vert\!\sim}_0 \rangle$ such that (i) $\mathscr{L}_0$ is a set of atomic sentences, $\{p_1, \ldots, p_n\} = \mathscr{L}_{0-}$, and the symbol $\bot$, and (ii) $\mathrel{\vert\!\sim}_0$ is a material consequence relation that also encodes an incoherence property, $\mathrel{\vert\!\sim}_0 \subseteq \mathcal{P}(\mathscr{L}_{0-}) \times \mathscr{L}_0$.*

Definition 3 *Proper Base Structure: A base structure is proper iff its underlying consequence relation, $\mathrel{\vert\!\sim}_0$, is proper, i.e., if it satisfies the following conditions: (a) $\mathscr{L}_{0-} \mathrel{\vert\!\sim}_0 \bot$, (b) $\emptyset \mathrel{\vert\!\not\sim}_0 \bot$, (c) $\mathrel{\vert\!\sim}_0$ is reflexive, and (d) ExFF.*

All base structures I will talk about are proper base structures. Our goal is to extend arbitrary proper base structures to structures with a language, $\mathscr{L}$, that contains logical vocabulary and a consequence relation, $\mathrel{\vert\!\sim}$, over this extended language. Moreover, this extension should be such that, firstly, we introduce logical expressions by giving rules that determine their roles in the extended consequence relation. This is the logical inferentialism. And secondly, the so introduced logical vocabulary should make explicit features of the consequence relation into which it is introduced. That is the logical expressivism. As already intimated, for our purposes, the second point amounts to two desiderata: the extended consequence relation should satisfy, the Deduction Theorem (DT) and what I shall call the "Negation Theorem" (NT):

DT $\Gamma \mathrel{\vert\!\sim} A \to B \iff \Gamma, A \mathrel{\vert\!\sim} B$.

NT $\Gamma \mathrel{\vert\!\sim} \neg A \iff \Gamma, A \mathrel{\vert\!\sim} \bot$.

If DT holds, the conditional is tracking the consequence relation. Such a conditional allows us to not only practically acknowledge that B follows from A in the context of Γ by inferring B from A in the context of Γ, but to assert something on the basis of Γ that commits us to B following from A, in the context of Γ. Similarly, if NT holds, the negation is tracking the incoherence property. Such a negation allows us to assert something on the basis of Γ, that commits us to Γ being incompatible with A (i.e., they being jointly incoherent). That is the sense in which such a conditional and negation make explicit the consequence relation and incoherence property of the language in which they occur.

There are two further desiderata for the extension of the underlying consequence relation. First, as explained above, we want the extension to be conservative, i.e., if $\Gamma \subseteq \mathscr{L}_{0-}$ and $A \in \mathscr{L}_0$, then $\Gamma \mathrel{\mid\!\sim} A$ iff $\Gamma \mathrel{\mid\!\sim}_0 A$. Second, we want the extension to preserve reflexivity, i.e., if the base consequence relation is reflexive, the extended one must be so, too.

Let's sum up the goal that I shall pursue in the reminder of this paper:

GOAL   We want to find a way to extend any proper base structure in such a way that the extension, $\langle \mathscr{L}, \mathrel{\mid\!\sim} \rangle$, is conservative, preserves reflexivity, and obeys DT and NT.

Notice that the conservativeness of the extension means that the extension must not force monotonicity. After all, a nonmonotonic consequence relation cannot be extended conservatively to a monotonic consequence relation.

As a bonus, I will also introduce a new modal operator, $\Box$. This operator marks consequences that hold monotonically. In order to see what this means, notice that even in a consequence relation where monotonicity fails as a global property, there can be sets of sentences, Γ, such that Γ and every superset of it imply a certain sentence A, i.e., $\forall \Delta \subseteq \mathscr{L}_- (\Delta, \Gamma \mathrel{\mid\!\sim} A)$. Thus, the implication $\Gamma \mathrel{\mid\!\sim} A$ behaves monotonically. I will introduce an operator that tracks this property of implications in the object language. More precisely, the operator will obey the following principle.

BOX $\Gamma \mathrel{\mid\!\sim} \Box A$ iff $\forall \Delta \subseteq \mathscr{L}_- (\Delta, \Gamma \mathrel{\mid\!\sim} A)$.

I will sometimes call this operator the "monotonicity-box." Having such an operator is not necessary for a logical system that exemplifies the ideas of logical inferentialism and logical expressivism in a nonmonotonic setting. In so far as regions where monotonicity holds locally are of interest, however, having such an operator is desirable.

3 The construction

In the previous section, I explained that we want to extend a material consequence relation to a language, $\mathscr{L}$, with logical vocabulary. The extended language I shall use includes negation, a conditional, conjunction, disjunction, and the new kind of modal operator mentioned in the previous section. The syntax of the language without $\bot$ is straightforward.

Syntax of $\mathscr{L}_-$: $\varphi ::= p \mid \neg\varphi \mid \varphi \to \varphi \mid \varphi \& \varphi \mid \varphi \vee \varphi \mid \Box\varphi$

And p is an atomic sentence of $\mathscr{L}_-$ iff $p \in \mathscr{L}_{0-}$. We now define the extended language as $\mathscr{L} = \mathscr{L}_- \cup \{\bot\}$.

Extending the consequence relation to $\mathrel{\mid\!\sim} \subseteq \mathcal{P}(\mathscr{L}_-) \times \mathscr{L}$ is more tricky. We do this by way of a sequent calculus in which the material implications serve as axioms. I call it the Non-Monotonic Modal sequent calculus (NMM).

We start with the straightforward idea that whatever is in the underlying consequence relation, $\mathrel{\mid\!\sim}_0$, is an axiom of the sequent calculus. However, there is a complication that has to do with our monotonicity-box. Our sequent calculus does not only have one kind of turnstile but $|\mathcal{P}(\mathcal{P}(\mathscr{L}_0))|$ many turnstiles. The idea is that, for every subset X of $\mathcal{P}(\mathscr{L}_0)$, we want to have $\Gamma \mathrel{\mid\!\sim}^{\uparrow X} A$ just in case $\forall \Delta \in X\,(\Delta, \Gamma \mathrel{\mid\!\sim} A)$. We stipulate this for the axioms of our sequent calculus and, hence, get axioms with different kinds of turnstiles.

Here is how we construct the extended consequence relation, $\mathrel{\mid\!\sim}$. First, we have two clauses that provide us with axioms of our sequent calculus.

Axioms of NMM:
Ax1: If $\Gamma \mathrel{\mid\!\sim}_0 A$, then $\Gamma \mathrel{\mid\!\sim} A$ is an axiom.
Ax2: If $X \subseteq \mathcal{P}(\mathscr{L}_0)$ and $\forall \Delta \in X\,(\Delta, \Gamma \mathrel{\mid\!\sim}_0 A)$, then $\Gamma \mathrel{\mid\!\sim}^{\uparrow X} A$ is an axiom.
Convention: If $X = \mathcal{P}(\mathscr{L}_0)$, we can write $\Gamma \mathrel{\mid\!\sim}^{\uparrow X} A$ as $\Gamma \mathrel{\mid\!\sim}^{\uparrow} A$.

A sequent is in $\mathrel{\mid\!\sim}$ just in case it can be derived from these axioms in a proof-tree using only the following sequent rules:

Rules of NMM:
Note on the notation: What is on the left of the turnstile are sets of formulae. The comma is to be read as set-union with flanking individual formulae being read as in set brackets; e.g., "Γ, A" on the left means "$\Gamma \cup \{A\}$". Upward arrows and formulae in square brackets are optional. That is, both, the sequent with and the sequent without the bracketed upward arrow, are derivable via the rule. Some rules are presented as involving ordinary sequents,

A Nonmonotonic Sequent Calculus

i.e. $\mathrel{\mid\!\sim}$-type sequents, but they also apply to quantified sequents, i.e. $\mathrel{\mid\!\sim}^{\uparrow X}$-type sequents. That is, they should be read as systematically ambiguous in the following way. Ordinary sequents can be replaced by quanitified sequents in unified ways. I.e. the rule can be applied if all the $\mathrel{\mid\!\sim}$-type turnstiles in the premises and the conclusion are replaced by $\mathrel{\mid\!\sim}^{\uparrow X}$-type turnstiles with the same X in all of these premises and the conclusion.

$$\frac{\Gamma, A \mathrel{\mid\!\sim} B}{\Gamma \mathrel{\mid\!\sim} A \to B} \text{ CP} \qquad \frac{\Gamma \mathrel{\mid\!\sim} A \to B}{\Gamma, A \mathrel{\mid\!\sim} B} \text{ CCP}$$

$$\frac{\Gamma, A_1, \ldots, A_n \mathrel{\mid\!\sim}^{\uparrow} B \quad \Gamma, B \mathrel{\mid\!\sim}^{\uparrow} A_1 \quad \ldots \quad \Gamma, B \mathrel{\mid\!\sim}^{\uparrow} A_n \quad \Gamma, C \mathrel{\mid\!\sim} D}{\Gamma, A_1, \ldots, A_n, B \to C \mathrel{\mid\!\sim} D} \text{ LC}$$

$$\frac{\Gamma, A \mathrel{\mid\!\sim} \bot}{\Gamma \mathrel{\mid\!\sim} \neg A} \text{ RN} \qquad \frac{\Gamma \mathrel{\mid\!\sim} A}{\Gamma, \neg A \mathrel{\mid\!\sim} \bot} \text{ LN}$$

$$\frac{\Gamma, A, B \mathrel{\mid\!\sim} C}{\Gamma, A\&B \mathrel{\mid\!\sim} C} \text{ L\&} \qquad \frac{\Gamma \mathrel{\mid\!\sim} A \quad \Gamma \mathrel{\mid\!\sim} B}{\Gamma \mathrel{\mid\!\sim} A\&B} \text{ R\&}$$

$$\frac{\Gamma, A \mathrel{\mid\!\sim} C \quad \Gamma, B \mathrel{\mid\!\sim} C}{\Gamma, A \vee B, [A], [B] \mathrel{\mid\!\sim} C} \text{ Lv} \qquad \frac{\Gamma \mathrel{\mid\!\sim} A}{\Gamma \mathrel{\mid\!\sim} A \vee B} \text{ Rv1} \qquad \frac{\Gamma \mathrel{\mid\!\sim} B}{\Gamma \mathrel{\mid\!\sim} A \vee B} \text{ Rv2}$$

$$\frac{\Gamma \mathrel{\mid\!\sim}^{\uparrow} A}{\Gamma \mathrel{\mid\!\sim} \Box A} \text{ RB} \qquad \frac{\Gamma, A \mathrel{\mid\!\sim} B}{\Gamma, \Box A \mathrel{\mid\!\sim} B} \text{ LB}$$

$$\frac{\Gamma \mathrel{\mid\!\sim}^{\uparrow} A}{\Gamma, B \to C \mathrel{\mid\!\sim} {}^{[\uparrow]}A} \text{ CK} \qquad \frac{\Gamma \mathrel{\mid\!\sim}^{\uparrow} A}{\Gamma, \neg B \mathrel{\mid\!\sim} {}^{[\uparrow]}A} \text{ NK} \qquad \frac{\Gamma \mathrel{\mid\!\sim}^{\uparrow} \bot}{\Gamma \mathrel{\mid\!\sim} {}^{[\uparrow]}A} \text{ ExFF}$$

$$\frac{\Gamma \mathrel{\mid\!\sim}^{\uparrow X} A \quad \Gamma \mathrel{\mid\!\sim}^{\uparrow Y} A}{\Gamma \mathrel{\mid\!\sim}^{\uparrow X \cup Y} A} \text{ UN} \qquad \frac{\Gamma, p_1 \ldots p_n \mathrel{\mid\!\sim} A}{\Gamma \mathrel{\mid\!\sim}^{\uparrow \{\{p_1 \ldots p_n\}\}} A} \text{ PushUp}$$

These sequent rules define a consequence relation $\mathrel{\mid\!\sim} \subseteq \mathcal{P}(\mathscr{L}_-) \times \mathscr{L}$. They also define many "quantified consequence relations" of the $\mathrel{\mid\!\sim}^{\uparrow X}$-type.

The purpose of the latter ones is merely auxiliary. They allow us to introduce the monotonicity-box, to use ExFF as a rule, and to use rules like LC, CK and NK for our conditional and negation. In this way, quantified sequents influence the extension of $\mathrel{\mid\!\sim}$ indirectly.

This construction gives us an extension of base structures: $\langle \mathscr{L}, \mathrel{\mid\!\sim} \rangle$. We now have to show that this extension satisfies the requirements set out in GOAL and BOX above.

4 Properties of the extension

Given GOAL and BOX above, we want the extended consequence relation to have the following properties:

1. $\mathrel{\mid\!\sim}$ is well defined.

2. $\mathrel{\mid\!\sim}$ is reflexive—i.e. $\forall \Gamma \subseteq \mathscr{L}_- (A \in \Gamma \Rightarrow \Gamma \mathrel{\mid\!\sim} A)$—if $\mathrel{\mid\!\sim}_0$ is reflexive.

3. $\mathrel{\mid\!\sim}$ is a conservative extension of $\mathrel{\mid\!\sim}_0$, i.e., for all $A \in \mathscr{L}_0$ and $\Gamma \subseteq \mathscr{L}_0$, $\Gamma \mathrel{\mid\!\sim}_0 A$ iff $\Gamma \mathrel{\mid\!\sim} A$.

4. DT holds, i.e., $\Gamma \mathrel{\mid\!\sim} A \to B$ iff $\Gamma, A \mathrel{\mid\!\sim} B$.

5. NT holds, i.e., $\Gamma \mathrel{\mid\!\sim} \neg A$ iff $\Gamma, A \mathrel{\mid\!\sim} \bot$.

6. BOX should hold, i.e., $\Gamma \mathrel{\mid\!\sim} \Box A$ iff $\forall \Delta \subseteq \mathscr{L}_- (\Gamma, \Delta \mathrel{\mid\!\sim} A)$.

Two remarks are in order: first, I restrict all these claims to finite premise sets; and I will assume that the base language is finite. I will not worry about compactness. This is a restriction of the current approach that I hope can be lifted for future descendants of it. Second, due to limitations of space I can only sketch the proofs of these properties. And sometimes I will omit proofs entirely.[4]

Restricting ourselves to finite premise sets, the first of these claims can easily be seen to be true because we only add sequents to our consequence relation. Since we never explicitly require something to *not* be in the relation, we cannot contradict ourselves. If we can show that our extension is conservative, this will also show that our consequence relation is not trivial, i.e., that it does not hold between every premise set and every formula. So let's look that conservativeness and the preservation of reflexivity.

---

[4]Contact me for detailed versions of the proofs.

A Nonmonotonic Sequent Calculus

4.1 Preservation of reflexivity and conservativeness

In order to show that reflexivity is preserved and that the extension is conservative, we first need some lemmas.

Lemma 1 *If* $p_1, \ldots, p_n, \Gamma \mathrel{|\!\sim} A$*, then* $\Gamma \mathrel{|\!\sim}^{\uparrow\{\{p_1 \ldots p_n\}\}} A$.

Proof. Immediate from PushUp. □

Lemma 2 *If* $\forall \Delta \subseteq \mathscr{L}_- (\Delta, \Gamma \mathrel{|\!\sim} A)$*, then* $\Gamma \mathrel{|\!\sim}^{\uparrow} A$.

Proof. Suppose that $\forall \Delta (\Delta, \Gamma \mathrel{|\!\sim} A)$. This implies $\forall \Delta \subseteq \mathscr{L}_{0-} (\Delta, \Gamma \mathrel{|\!\sim} A)$. So, for every subset of our atoms, $\{p_1 \ldots p_m\}$, we have $p_1 \ldots p_m, \Gamma \mathrel{|\!\sim} A$. So, by lemma 1, $\Gamma \mathrel{|\!\sim}^{\uparrow\{\{p_1 \ldots p_m\}\}} A$. By $2^{|\mathscr{L}_0|}$ applications of UN, we get $\Gamma \mathrel{|\!\sim}^{\uparrow} A$. □

Next, we need a lemma that says that if we can weaken a sequent with arbitrary sets of atoms, then we can weaken it with arbitrary sets of formulae.

Lemma 3 *If* $\forall \Delta \subseteq \mathscr{L}_{0-} (\Delta, \Gamma \mathrel{|\!\sim} A)$*, then* $\forall \Delta \subseteq \mathscr{L}_- (\Delta, \Gamma \mathrel{|\!\sim} A)$.

Proof. By induction on the complexity of the most complex formulae in Δ, where complexity is the number of connectives in a formula. The base case is immediate. For the induction step, take an arbitrary set, Θ, with the maximally complex formulae in it being of complexity $n+1$. We divide Θ into the following sets: N is the set of formulae of complexity $\leq n$, C is the set of conditionals of complexity $n+1$, *NEG* is the set of negations of complexity $n+1$, *CON* is the set of conjunctions of complexity $n+1$, D is the set of disjunctions of complexity $n+1$, and B is the set of necessitations of complexity $n+1$. So, $\Theta = N \cup C \cup NEG \cup CON \cup D \cup B$. Looking at the proof of lemma 2 again, we know that the antecedent of our conditional gives us $\Gamma \mathrel{|\!\sim}^{\uparrow} A$. So we can easily weaken with N, C, and *NEG*. We can also weaken with the embedded formulae of conjunctions, disjunctions and necessitations of complexity n. From this we can derive the conjunctions and necessitations via L& and LB. So the only potential difficulty is weakening with disjunctions of complexity n. In order to do this, we make a list of all the formulae that are the disjuncts of the k elements of D: $d_{1.1}, d_{1.2}, d_{2.1} \ldots d_{k-1.2}, d_{k.1}, d_{k.2}$, where the first index indicates the number of the disjunction from which the formula stems and the second index indicates whether it is the first or the second disjunct. We take the 2^k different subsets from this list for which: for each $1 \leq n \leq k$ exactly one of $d_{n.1}$ or $d_{n.2}$ is in the set and nothing else is in the set. Call these sets $\Xi_1 \ldots \Xi_{2^k}$. Let $\Pi = N \cup C \cup NEG \cup CON \cup B \cup \Gamma$. Thus, for each $1 \leq m \leq 2^k$, we get $\Xi_m, \Pi \mathrel{\mid\!\sim} A$. We now construct our proof-tree in the following way:

$$\dfrac{\dfrac{\ldots \mathrel{\mid\!\sim} \ldots \quad \ldots \mathrel{\mid\!\sim} \ldots}{d_{1.1} \vee d_{1.2} \ldots d_{k.1}, \Pi \mathrel{\mid\!\sim} A} \text{Lv} \quad \dfrac{\ldots \mathrel{\mid\!\sim} \ldots \quad \ldots \mathrel{\mid\!\sim} \ldots}{d_{1.1} \vee d_{1.2} \ldots d_{k.2}, \Pi \mathrel{\mid\!\sim} A} \text{Lv}}{d_{1.1} \vee d_{1.2} \ldots d_{k.1} \vee d_{k.2}, \Pi \mathrel{\mid\!\sim} A} \text{Lv}$$

Since Θ was arbitrary, we have $\forall \Delta (\Delta, \Gamma \mathrel{\mid\!\sim} A)$ for Δs of complexity $n+1$. $\square$

Proposition 1 *The extension preserves reflexivity.*

Proof. We assume that $\mathrel{\mid\!\sim}_0$ is reflexive. First, we show, by induction on the complexity of α, that $\forall \Delta \subseteq \mathscr{L}_{0-} (\Delta, \alpha \mathrel{\mid\!\sim} \alpha)$. And by lemma 3, this implies that $\forall \Delta \subseteq \mathscr{L}_- (\Delta, \alpha \mathrel{\mid\!\sim} \alpha)$. $\square$

We now know that the first two of the six points above hold. Before we turn to conservativeness, we need two more lemmas.

Lemma 4 *If $\Gamma \mathrel{\mid\!\sim}^{\uparrow X} A$, then $\forall \Delta \in X (\Delta, \Gamma \mathrel{\mid\!\sim} A)$.*

Proof. By induction on proof height, i.e., the number of rule-applications in the longest branch of the proof-tree. The proof is, for the most part, straightforward. I will leave some minor complications with LC, UN and PushUp as an exercise for the reader. $\square$

Lemmas 3 and 4, when applied to the case where $X = \mathcal{P}(\mathscr{L}_0)$, imply the following:

Lemma 5 *If $\Gamma \mathrel{\mid\!\sim}^\uparrow A$, then $\forall \Delta \subseteq \mathscr{L}_- (\Delta, \Gamma \mathrel{\mid\!\sim} A)$.*

With these lemmas in hand, we can show that the extension is conservative for any underlying consequence relation that obeys ExFF.

Proposition 2 *The extension is a conservative extension of any non-monotonic material consequence relation that obeys ExFF; i.e., for all $A \in \mathscr{L}_0$ and $\Gamma \subseteq \mathscr{L}_0$, $\Gamma \mathrel{\mid\!\sim}_0 A$ iff $\Gamma \mathrel{\mid\!\sim} A$.*

A Nonmonotonic Sequent Calculus

Proof. The left-to-right direction is immediate from Ax1. So we only have to show that our construction does not add any sequent that can be formulated in the base language and is not already in $\mathrel{\mid\!\sim}_0$. We argue by *reductio*, and we look at the (or a) shortest possible proof of a given violation of conservativeness (where length is the number of rule applications in a prooftree). If NMM allows a violation of conservativeness, the last step is either an application of CCP or of ExFF. After all, the other rules have logical connectives in the conclusion-sequent; or else they apply only to quantified sequents. So we have two cases:

(Case 1) Assume that the violation, $\Gamma \mathrel{\mid\!\sim} p$, comes by ExFF. The premise is $\Gamma \mathrel{\mid\!\sim}^{\uparrow} \bot$. This must come by Ax2 or by UN. If it comes by Ax2, we have $\forall \Delta\, (\Delta, \Gamma \mathrel{\mid\!\sim}_0 \bot)$. But $\mathrel{\mid\!\sim}_0$ obeys ExFF by stipulation. So $\Gamma \mathrel{\mid\!\sim} p$ cannot violate conservativeness. Hence, $\Gamma \mathrel{\mid\!\sim} p$ must come by UN. The premises are $\Gamma \mathrel{\mid\!\sim}^{\uparrow X} \bot$ and $\Gamma \mathrel{\mid\!\sim}^{\uparrow Y} \bot$ and $X \cup Y = \mathcal{P}(\mathscr{L}_{0-})$. It can be shown by induction on proof height that if Γ contains only atoms and $\Gamma \mathrel{\mid\!\sim}^{\uparrow X} \bot$, then $\forall \Delta \in X\, (\Delta, \Gamma \mathrel{\mid\!\sim}_0 \bot)$. Thus, we get $\forall \Delta \in \mathcal{P}(\mathscr{L}_{0-})\, (\Delta, \Gamma \mathrel{\mid\!\sim}_0 \bot)$. And by ExFF for the underlying consequence relation we have $\Gamma \mathrel{\mid\!\sim}_0 p$.

(Case 2) Assume that the violation comes by CCP. The premise is $\Gamma \mathrel{\mid\!\sim} A \to B$. This must come by CP or CCP or ExFF. If it comes by CP, the premise is $\Gamma, A \mathrel{\mid\!\sim} B$. This violates our assumption that there is no shorter proof of $\Gamma, A \mathrel{\mid\!\sim} B$. So, it must come by CCP or ExFF.

(Case 2.a) If $\Gamma \mathrel{\mid\!\sim} A \to B$ comes by CCP, the premise is $\Gamma \setminus \{C\} \mathrel{\mid\!\sim} C \to (A \to B)$. Since $\Gamma \setminus \{C\}$ contains only atoms, we are in the same situation again: either it must come by CP, CCP, or ExFF. If it comes by CP, we are back at $\Gamma \mathrel{\mid\!\sim} A \to B$. If it comes by CCP, the premise is $\Gamma \setminus \{C, D\} \mathrel{\mid\!\sim} D \to (C \to (A \to B))$. The same question arises again. If we continue like that, we are launched on an infinite regress of CCP applications. So at some point the conditional must come by ExFF. But if one of these conditionals comes by ExFF some subset, Θ, of Γ must be persistently incoherent, i.e., $\Theta \mathrel{\mid\!\sim}^{\uparrow} \bot$. By lemma 5, $\forall \Delta \subseteq \mathscr{L}_{-}\, (\Delta, \Theta \mathrel{\mid\!\sim} \bot)$. Since ExFF cannot conclude a violation of conservativeness (see Case 1) and everything in Θ is atomic, we have $\forall \Delta\, (\Delta, \Theta \mathrel{\mid\!\sim}_0 \bot)$. Hence, $\forall \Delta\, (\Delta, \Gamma, A \mathrel{\mid\!\sim}_0 \bot)$. But then ExFF for $\mathrel{\mid\!\sim}_0$ applies.

(Case 2.b) $\Gamma \mathrel{\mid\!\sim} A \to B$ comes by ExFF. The same reasoning as in the previous subcase applies. □

We now know that our extension is well-defined, conservative and that it preserves reflexivity. So we can now turn to the last three properties listed at the beginning of this section.

4.2 Behavior of the conditional, negation, and box

We want the conditional to express the consequence relation, the negation to express incoherence, and the box to express monotonicity. What this comes to, for our purposes, is that DT, NT, and BOX hold.

It is immediate that DT holds. After all, CP gives us the right-to-left direction, and CCP gives us the left-to-right-direction. Parenthetically, it is worth pointing out the CCP is a simplifying rule. This would lead to problems if we wanted to prove Cut-elimination. As I will explain below, however, we don't want to do that.

Regarding negation, we want NT to hold, i.e., we want:

Proposition 3 $\Gamma, A \mathrel{\mid\!\sim} \bot \Leftrightarrow \Gamma \mathrel{\mid\!\sim} \neg A$.

Proof. The left-to-right direction is immediate because we have RN. So we must show that if $\Gamma \mathrel{\mid\!\sim} \neg A$, then $\Gamma, A \mathrel{\mid\!\sim} \bot$. We argue by induction on the height of a shortest proof of $\Gamma \mathrel{\mid\!\sim} \neg A$. Base case: $\Gamma \mathrel{\mid\!\sim} \neg A$ comes by the application of just one rule. It must come by RN or ExFF. In either case, we have $\Gamma, A \mathrel{\mid\!\sim} \bot$. Induction step: our hypothesis is that if $\Gamma \mathrel{\mid\!\sim} \neg A$ can be derived in a proof of height n, then $\Gamma, A \mathrel{\mid\!\sim} \bot$. For a proof of height $n+1$, the last rule applied can be: CCP, RN, L&, Lv, CK, NK, LC, LB, or ExFF. It is easy to see that in the cases of RN, ExFF, L&, Lv, CK, NK, and LB we get $\Gamma, A \mathrel{\mid\!\sim} \bot$ in one or two steps. For, either the premise itself is $\Gamma, A \mathrel{\mid\!\sim} \bot$, or we apply the hypothesis to the premise and derive $\Gamma, A \mathrel{\mid\!\sim} \bot$ with the same rule, or we get it by ExFF. So we are left with two cases.

(Case 1) the last rule applied is CCP. The premise is $\Gamma \setminus \{B\} \mathrel{\mid\!\sim} B \to \neg A$. If this comes by CP or ExFF, $\Gamma, A \mathrel{\mid\!\sim} \bot$ is immediate. If it comes by CCP, L&, Lv, CK, NK, LC, or LB, this also gives us what we want. As an example, suppose it comes by L&. The premise is $\Gamma \setminus \{B, C\&D\}, C, D \mathrel{\mid\!\sim} B \to \neg A$. We can argue thus:

$$\dfrac{\dfrac{\dfrac{\Gamma \setminus \{B, C\&D\}, C, D \mathrel{\mid\!\sim} B \to \neg A}{\Gamma \setminus \{C\&D\}, C, D \mathrel{\mid\!\sim} \neg A} \text{CCP}}{\Gamma \setminus \{C\&D\}, C, D, A \mathrel{\mid\!\sim} \bot} \text{Hyp}}{\Gamma, A \mathrel{\mid\!\sim} \bot} \text{L\&}$$

Lv, CK, NK, and LB work in an analogous way.

Next suppose $\Gamma \setminus \{B\} \mathrel{\mid\!\sim} B \to \neg A$ comes by LC. The right premise is $\Gamma \setminus \{C_1, \ldots, C_n, D \to E\}, E \mathrel{\mid\!\sim} B \to \neg A$. So, by our hypothesis, $A, B, \Gamma \setminus \{C_1, \ldots, C_n, D \to E\}, E \mathrel{\mid\!\sim} \bot$. The other premises are: $\Gamma \setminus \{D \to$

$E\}, E \mathrel{|\!\sim^\uparrow} D$, and $\Gamma \setminus \{D \to E\}, D \mathrel{|\!\sim^\uparrow} C_1$, and …, and $\Gamma \setminus \{D \to E\}, D \mathrel{|\!\sim^\uparrow} C_n$. By lemma 5, the upward arrow implies that we can weaken with $\{A, B\}$. Hence, $A, B, \Gamma \setminus \{D \to E\}, E \mathrel{|\!\sim^\uparrow} D$, and $A, B, \Gamma \setminus \{D \to E\}, D \mathrel{|\!\sim^\uparrow} C_1$, and …, and $A, B, \Gamma \setminus \{D \to E\}, D \mathrel{|\!\sim^\uparrow} C_n$. So by LC, $\Gamma, A \mathrel{|\!\sim} \bot$.

Suppose $\Gamma \setminus \{B\} \mathrel{|\!\sim} B \to \neg A$ comes by CCP. The premise is $\Gamma \setminus \{B, C\} \mathrel{|\!\sim} C \to (B \to \neg A)$. The reasoning we just went through applies again. So CCP cannot conclude a sequent that contradicts our proposition.

(Case 2) the last rule applied is LC. We apply the same reasoning that we applied in the LC subcase of (Case 1). $\square$

Finally, we must show that BOX holds. We divide BOX into two parts:

- $\Gamma \mathrel{|\!\sim} \Box A$ iff $\Gamma \mathrel{|\!\sim^\uparrow} A$.

- $\Gamma \mathrel{|\!\sim^\uparrow} A$ iff $\forall \Delta \, (\Delta, \Gamma \mathrel{|\!\sim} A)$.

Regarding the second part of BOX, notice that we have already proven both directions of this principle as lemmas 2 and 5. So we already know that:

Proposition 4 $\forall \Delta \, (\Delta, \Gamma \mathrel{|\!\sim} A)$ *iff* $\Gamma \mathrel{|\!\sim^\uparrow} A$.

Hence, it is just the first part of BOX that still needs to be proven. In order to do so, we again first need a lemma.

Lemma 6 *If* $\Gamma \mathrel{|\!\sim} B_1 \to (B_2 \ldots \to (B_k \to \Box A))$, *then* $\Gamma, B_1 \ldots B_k \mathrel{|\!\sim^\uparrow} A$.

Proof. By induction on proof height of $\Gamma \mathrel{|\!\sim} B_1 \to (B_2 \ldots \to (B_k \to \Box A))$. The only tricky case is the induction step for LC. It goes as follows: The premises are $\Gamma \setminus \{D \to E\} \mathrel{|\!\sim^\uparrow} D$, and $\Gamma \setminus \{C_1, \ldots, C_n, D \to E\}, D \mathrel{|\!\sim^\uparrow} C_1, \ldots, \Gamma \setminus \{C_1, \ldots, C_n, D \to E\}, D \mathrel{|\!\sim^\uparrow} C_n$, and $\Gamma \setminus \{C_1, \ldots, C_n, D \to E\}, E \mathrel{|\!\sim} B_1 \to (B_2 \ldots \to (B_k \to \Box A))$. By our hypothesis, $\Gamma \setminus \{C_1, \ldots, C_n, D \to E\}, E, B_1 \ldots B_k \mathrel{|\!\sim^\uparrow} A$. By a couple of CP application, this gives us $\Gamma \setminus \{C_1, \ldots, C_n, D \to E\}, E \mathrel{|\!\sim^\uparrow} B_1 \to (B_2 \ldots \to (B_k \to A))$. Together with the other premises, LC allows us derive: $\Gamma \mathrel{|\!\sim^\uparrow} B_1 \to (B_2 \ldots \to (B_k \to A))$. And by iterated CCP, $\Gamma, B_1 \ldots B_k \mathrel{|\!\sim^\uparrow} A$. $\square$

We can now prove the first part of our BOX-principle.

Proposition 5 $\Gamma \mathrel{|\!\sim} \Box A$ *iff* $\Gamma \mathrel{|\!\sim^\uparrow} A$.

Proof. First, the left-to-right direction. We argue by induction on proof height. Base case: The shortest proof-tree of such a sequent is RB or ExFF and both guarantee that $\Gamma \mathrel{|\!\sim^\uparrow} A$. For the induction step, notice that the last step in a proof-tree for $\Gamma \mathrel{|\!\sim} \Box A$ can be CCP, L&, Lv, CK, NK, LC, RB, LB, or ExFF. Lemma 6 gives us the induction step for CCP. The others are straightforward and I'll leave them as an exercise for the reader. The right-to-left direction is immediate because of RB. □

From propositions 5 and 4 the desired BOX follows immediately. Thus, we have shown that the extension defined by our sequent rules has all the six properties we want it to have. Hence, we have a sequent calculus that is in line with logical inferentialism and logical expressivism.

4.3 Why does cut fail?

Before I move on to the relation between NMM and intuitionism, I want to point out a feature of the system that might seem problematic: the consequence relation $\mathrel{|\!\sim}$ is not transitive. That is, Cut is not only not provable but it actually fails. Monotonicity, transitivity, and reflexivity are often considered essential to anything being a consequence relation. Of course, we already abandoned that idea when we started to do nonmonotonic logic. But that we are also giving up transitivity might seem like a problem. I don't think it is a problem.[5] Rather, it is an insight that if you want to have a conditional that obeys a deduction theorem in a nonmonotonic setting, you need to give up transitivity.

To see this, take a mixed context version of Cut (Cut-MC):

$$\frac{\Gamma, A \mathrel{|\!\sim} B \quad \Delta \mathrel{|\!\sim} A}{\Gamma, \Delta \mathrel{|\!\sim} B} \text{Cut-MC}$$

Proposition 6 *Cut-MC together with reflexivity implies that if* $\Gamma \mathrel{|\!\sim} A$, *then* $\Gamma, \Delta \mathrel{|\!\sim} A$.

Proof. We argue thus:

$$\frac{\Gamma, \Delta, A \mathrel{|\!\sim} A \quad \Gamma \mathrel{|\!\sim} A}{\Gamma, \Delta \mathrel{|\!\sim} A} \text{Cut-MC}$$

The left premise is an instance of reflexivity and, hence, can be derived. □

---

[5] Dave Ripley has provided some independent motivation to be skeptical about Cut (Ripley, 2013, 2015; see also Schroeder-Heister, 2004).

So you cannot have a mixed context version of Cut in a nonmonotonic system with a reflexive consequence relation.

One might move to a shared context version of Cut (Cut-SC) to get rid of this problem.

$$\frac{\Gamma, A \mathrel{|\!\sim} B \quad \Gamma \mathrel{|\!\sim} A}{\Gamma \mathrel{|\!\sim} B} \text{ Cut-SC}$$

However, if we have a deduction theorem, we can run a similar argument for monotonicity with Cut-SC:

$$\frac{\dfrac{\dfrac{\Gamma, A, B \mathrel{|\!\sim} A}{\Gamma, A \mathrel{|\!\sim} B \to A} \text{ CP} \quad \Gamma \mathrel{|\!\sim} A}{\Gamma \mathrel{|\!\sim} B \to A} \text{ Cut-SC}}{\Gamma, B \mathrel{|\!\sim} A} \text{ CCP}$$

Hence, if you want a nonmonotonic, reflexive consequence relation with a conditional that obeys a deduction theorem, you need to give up Cut—even the shared context version. Of course, you can reason by modus tollens at this point (see Morgan, 2000). I think, however, that given the plausibility of nonmonotonicity and reflexivity and the logical expressivist motivation for DT, there is good reason to at least investigate systems in which Cut fails along with monotonicity.

There may be particularly well behaved regions of logical space in which transitivity holds. And in the fullness of time, we hope to study such regions systematically and perhaps even to introduce an object language operator that lets us mark such regions. Here I just want to point out that the failure of Cut is not an unmotivated quirk of the NMM system. It is entailed by the properties that I require the system to have.

5 Relation of NMM to intuitionistic and classical logic

I want to briefly describe the surprisingly straightforward relation between NMM and intuitionistic and classical logic. Due to limitations of space, I will omit the proofs of the results I am presenting.

I have already pointed out that Cut-SC fails in NMM. However, if we add Cut-SC to our sequent rules, the NMM rules are equivalent to Gentzen's sequent rules for intuitionistic logic, LJ, modulo the rules governing the box (which is pointless in a monotonic system). Call the system that results from adding Cut-SC to the NMM rules the "Cut-System." Moreover, read a sequent with $\bot$ on the right in the Cut-System as meaning the same as a sequent with an empty right side in Gentzen's LJ. Translate all other sequents in the obvious way. It can be shown that, under this translation, the following holds:

Proposition 7 *Translations of all rules of Gentzen's LJ system can be derived in the Cut-System, and translations of all rules of the Cut-System that don't use the box or sequents quantifying over less than $\mathcal{P}(\mathcal{L}_0)$ can be derived in Gentzen's LJ.*

In effect, the Cut-System without the apparatus governing the box is equivalent to Gentzen's LJ. This does not only hold at the level of sequent rules, but also at the level of theorems. All the theorems of intuitionistic logic are theorems of the Cut-System, i.e., they are implied by the empty set. Given these facts, it is easy to see that the following holds:

Proposition 8 *If the underlying material consequence relation contains all and only instances of reflexivity and we ignore the box (by deleting RB and LB), the (non-quantified) consequence relation of the Cut-System coincides with the intuitionistic consequence relation.*

Since the Cut-System gives us intuitionism, it is clear that adding double negation elimiation to the Cut-System will give us classical logic. It is easy to add sequent rules that give us double negation elimination. Hence, classical logic can be recovered by adding Cut and such further sequent rules to NMM. In this sense, the system I have presented can be viewed as a "mother-logic" that gives rise to intuitionism or classical logic under special circumstances.

6 Conclusion

I have presented a way of extending a nonmonotonic material consequence relation over an atomic base language to a consequence relation over a logically complex language. The extension is conservative; it preserves reflexivity; the conditional tracks the consequence relation; the negation tracks the incoherence property; and a new kind of modal operator tracks local regions of monotonicity. Thus, I have presented a logical system that does justice to the philosophical ideas of logical inferentialism and logical expressivism.

References

Antonelli, G. A. (2008). Non-monotonic Logic. In E. N. Zalta (Ed.), *Stanford Encyclopedia of Philosophy* (Fall 2015 ed.). Retrieved from `http://plato.stanford.edu/entries/logic-nonmonotonic/`

Belnap, N. (1962). Tonk, Plonk and Plink. *Analysis, 22*(6), 130–134.

Brandom, R. B. (2000). *Articulating Reasons: an Introduction to Inferentialism.* Cambridge, Massachusetts: Harvard University Press.

Brandom, R. B. (2008). *Between Saying and Doing: Towards an Analytic Pragmatism.* Oxford: Oxford University Press.

Brandom, R. B. (2009). *Reason in Philosophy: Animating Ideas.* Cambridge, Massachusetts: Harvard University Press.

Dummett, M. A. E. (1991). *The Logical Basis of Metaphysics.* Cambridge, Massachusetts: Harvard University Press.

Gentzen, G. (1934). Untersuchungen über das Logische Schließen: I. *Mathematische Zeitschrift, 39*(2), 176–210.

Kneale, W. C. (1956). The Province of Logic. In H. Lewis (Ed.), *Contemporary British Philosophy* (pp. 235–261). London: Allen & Unwin.

Morgan, C. G. (2000). The Nature of Nonmonotonic Reasoning. *Minds and Machines, 10*(3), 321–360.

Peregrin, J. (2014). *Inferentialism: Why Rules Matter.* New York City: Palgrave MacMillan.

Prior, A. N. (1960). The Runabout Inference-ticket. *Analysis, 21*(2), 38–39.

Ripley, D. (2013). Paradoxes and Failures of Cut. *Australasian Journal of Philosophy, 91*(1), 139–164.

Ripley, D. (2015). Naive Set Theory and Nontransitive Logic. *Review of Symbolic Logic, 8*(3), 553–571.

Schroeder-Heister, P. (2004). On the Notion of 'Assumption' in Logical Systems. In R. Bluhm & C. Nimtz (Eds.), *Fifth International Congress of the Society for Analytical Philosophy, Bielefeld, 22–26 September 2003* (pp. 27–48). Paderborn: Mentis.

Ulf Hlobil
University of Pittsburgh
USA
E-mail: `ulh1@pitt.edu`

The Larger Logical Picture

JOHN T. KEARNS

Abstract: In this paper, I articulate a conceptual framework which accommodates speech acts, or language acts, and logical theories. This assists in identifying topics and areas that can be, and should be, explored. Many have been explored already, but many others have not. Developing the framework is preliminary to working out theories for the unexplored parts, because the framework itself furnishes guidance for doing this. In John Searle's taxonomy of illocutionary acts, the three categories assertive, directive, and commissive are the basis for three fundamental types of logical theory. For each category of illocutionary acts, there are characteristic locutionary acts, and three types of deductive argument, or derivation: locutionary arguments, deductive derivations, and illocutionary arguments. Most familiar logical theories deal with assertive locutionary acts and locutionary arguments. More attention needs to be paid to directive and commissive logical theories, and to illocutionary arguments associated with all three categories.

Keywords: speech acts, language acts, illocutionary logic

1 Illocutionary acts

In this paper, I am concerned to articulate a conceptual framework which accommodates speech acts, or language acts, as well as logical theories. I will sketch a landscape which identifies topics and areas to be explored. Some of these have been explored already, but many have not. Developing this framework is preliminary to working out theories for the unexplored parts, because the framework itself furnishes guidance for doing this.

A language act is a meaningful act performed by saying something, or writing something, or even thinking something with words and sentences. *Illocutionary acts* are the fundamental kind of language act. A typical illocutionary act is performed by someone's using a sentence or a sentential clause to say something meaningful, and doing this in a certain manner or with a certain force, like the force of an assertion, a request, or a promise. Illocutionary acts are the complete concrete language acts of which significant speech is composed.

John T. Kearns

John Searle's taxonomy of illocutionary acts (introduced in Searle, 1985) recognizes five major categories of these acts, and three of these categories are relevant to the present paper. These are *assertives*, *directives*, and *commissives*. Mike's assertion that $\sqrt{2}$ is a rational number is an *assertive illocutionary act*. So would be his denial that $\sqrt{2}$ is rational, or his act of supposing that $\sqrt{2}$ is rational, or his act of supposing that $\sqrt{2}$ is not rational. What is asserted, denied, or supposed in these illocutionary acts is the *statement* that $\sqrt{2}$ is rational. A statement is an act performed with a sentence or sentential clause that can appropriately be evaluated in terms of truth and falsity. Statements are *locutionary acts*. Performing a statement with the force of an assertion would constitute an assertion.

In a *directive illocutionary act*, a speaker or writer tries to get her addressee to do or not do something. She might, for example, order her addressee to shut the door, or ask her addressee to shut the door, or advise him to shut the door. The locutionary act that is abstracted from a directive act, and that is performed with a specific force to constitute the directive act is a *plan* for the addressee to *implement*.

A plan is a speech act like "Mark, please close the door" or "Rosemary, answer the phone," and represents the addressee as performing the directed act. Just as statements have truth conditions which are satisfied or not, so plans have *implementation conditions* for which we will also speak of satisfaction. If a directive act has the desired effect on its addressee, that addressee will, first, *agree* to implement the directive's plan, and, subsequently, will implement the plan.

A directive illocutionary act is an act performed by a speaker to get her addressee to implement a plan, after that addressee first commits himself to do this. *A commissive illocutionary act* is also concerned with implementing a plan. In the commissive case, the speaker commits *herself* to implement a plan. For example, a person might say, or think, "I will get a bottle of beer from the refrigerator." Or she might say to her friend, "Rosemary, I will meet you for lunch on Thursday, at one o'clock at Lombardo's restaurant. I'm paying, and that's a promise."

Commissives resemble directives in having locutionary components that are plans, although the commissive plans are first-person while the directive plans are second-person. Commissives resemble assertives in that both kinds of act may have addressees, but addressees are not required. Someone can make an assertion or a decision when she is alone. But addressees are essential for directive acts.

The Larger Logical Picture

Searle (1969, and elsewhere) has argued for the non-importance of locutionary acts, either because they don't exist or because they are mere abstractions as opposed to the complete concrete illocutionary acts. Locutionary acts certainly exist, as I illustrated by writing "$\sqrt{2}$ is a rational number." But Searle is right that, ordinarily, locutionary acts are abstract components of illocutionary acts. However, this doesn't mean that locutionary acts are of no importance. For example, statements and features of statements are the focus of attention for most standard theories of deductive logic.

Actual language acts are performed by particular people on particular occasions. But different people can perform language acts that are essentially similar to one another. This makes it convenient to consider those different people to be performing the same acts, as when we say that each of us can make the statement that $\sqrt{2}$ is rational. But while different people can make the same statements, they can't perform the same illocutionary acts. Each person's illocutionary acts are uniquely her own.

2 Arguments and derivations

Three kinds of argument or derivation are associated with each of the three categories of illocutionary acts. To explain these, I will begin with assertive acts. Assertions, denials, and suppositions are fundamental kinds of assertive acts. Someone's *assertion* is constituted by her making a statement and accepting that statement as representing what is the case. Denials reject statements for failing to represent what is the case, and suppositions either temporarily accept statements or temporarily reject them. A person makes suppositions in constructing an argument, and suppositions are frequently discharged in the course of that argument.

An assertive *locutionary argument* is an ordered pair whose first member is a set of statements, the premises, and whose second member is the conclusion statement. Locutionary arguments are abstractions which we can represent and evaluate, but they are not arguments that a person can actually make or address to someone else. An assertive locutionary argument is *valid* iff its premises entail or imply the conclusion.

It is common to show that a locutionary argument is valid by constructing a *deductive derivation* tracing truth conditional connections from some premises to the conclusion of the argument. Although such derivations are often called arguments, or proofs, I prefer the term 'derivation.' Locutionary arguments and deductive derivations are distinct from *illocutionary ar-*

guments, in which a person reasons from premises which are illocutionary acts to an illocutionary act conclusion. Assertive illocutionary arguments are the "real life" arguments that someone uses to explore or extend her own knowledge and belief, or to convince someone else to accept or deny a given statement. In spite of this, standard accounts of deductive logic are almost exclusively concerned with locutionary arguments and deductive derivations. However, the systems presented in (Kearns, 2006, 2009a, 2009b) do explore assertive illocutionary arguments.

Deductive illocutionary arguments are based on *rational commitment*: a simple assertive illocutionary argument is *deductively correct* if performing the premiss acts commits the arguer to perform the conclusion act. A complex assertive illocutionary argument is deductively correct if its component arguments are deductively correct, and its initial undischarged premiss acts commit the arguer to perform the conclusion act.

Rational commitment is a person's commitment to do or not do something, or to remain in a certain state like that of accepting a given statement. Deciding to perform an act will commit a person to perform that act, and sometimes performing one intentional act commits a person to perform another. This commitment is either conditional or unconditional, and is either immediate or mediate; what makes someone's commitment immediate is that it is evident to her if she gives the matter some thought. Rational commitment is discussed, and explained, more fully in (Kearns, 2006, 2007, 2009a, 2009b).

The *inferential* commitment characteristic of assertive illocutionary acts is not a commitment to carry out reasoning, but is instead a commitment to recognize and follow certain links *when* carrying out deductive reasoning. My assertion that today is Wednesday inferentially commits me to assert that the day after tomorrow is Friday. The Wednesday assertion doesn't require me to give any thought to the following days, but my assertion commits me to grant, or concede, that the day after tomorrow is Friday, if the matter comes up.

3 Directive arguments

The logical theory of directive acts is sometimes called imperative logic, but it is misleading to label the theory with the name of one of its proper parts. It is the truth or satisfaction conditions of statements which provides a basis for speaking of entailment or implication linking some statements to others.

The Larger Logical Picture

In an analogous fashion, the implementation or satisfaction conditions of plans allow us to speak of entailment or implication linking some plans to others.

A set of plans for a single addressee *weakly entails* a further plan for that same addressee if any way of implementing the first plans will also implement the further plan. This is not such an interesting relation, because if Mark has been asked to *mail a letter*, and he complies, he has also implemented the plan "Mark, mail this letter or burn it," although he has *not* been asked to implement the disjunctive plan.

A plan, "Mark, do F" *weakly entails* the plan "Mark, do F or do G." To have "Mark, do F" strongly entailing "Mark, do G," we need the first plan to weakly entail the second, and, additionally, we need implementing the second plan to be part of what is involved in implementing the first. For example, "Mark, get up from your seat and shut the door" strongly entails "Mark, get up from your seat." One project for directive logic is to adequately characterize or explain strong entailment.

Since plans and sets of plans can entail other plans, there are locutionary arguments from plan premisses to plans as conclusions. Consider this directive locutionary argument:

< {Mark, get up from your seat and close the door}, Mark, get up from your seat. >

(The ordered pair notation is what marks this as a locutionary argument.) This is valid, because the premiss strongly entails the conclusion. But this is not an argument addressed to Mark. The premiss does not give Mark a reason to implement the conclusion. We focus on arguments like this in order to investigate entailment relations linking some plans to others.

We should notice that the premisses of directive locutionary arguments can include statements as well as plans. The following is an example:

< {Mark, if it rains, close the windows, It is raining}, Mark, close the windows. >

To evaluate this argument, we need to consider both truth conditions and implementation conditions, and we need to provide an account of conditional plans and their impact on implementation. Since truth conditions and implementation conditions are both *satisfaction conditions*, we should probably define validity in terms of satisfaction.

This most recent locutionary argument might easily be confused with an argument whose premises give Mark a reason to implement the conclusion. But that would be to regard the locutionary argument as an illocutionary argument. As well as considering directive locutionary arguments, it should also be possible, though perhaps not very interesting, to carry out deductive derivations tracing truth and implementation condition connections linking premisses of directive locutionary arguments to their conclusions.

If an assertive locutionary argument is valid, and someone knows or believes the premisses of that argument, then that person should be able to construct a deductively correct assertive illocutionary argument from assertions of the locutionary argument's premisses to the assertion of that argument's conclusion. The situation is different with directive arguments. A speaker cannot so easily transform a strongly valid directive locutionary argument like this:

< {Rachel, water the lawn every day next week}, Rachel, water the lawn next Thursday. >

into a deductively correct directive illocutionary argument. For such an argument should begin with assertions, and make clear to the addressee who accepts the asserted statements that she is already committed to implement the conclusion's plan. An example is the following:

Rachel, you have agreed to water the lawn every day next week.
So be sure to water the lawn next Thursday.

The premiss is an assertion, and the conclusion is a directive illocutionary act.

Directive illocutionary arguments are intended to give their addressees reasons to implement their conclusions' plans. The strongest reasons from a logical point of view are provided by deductive arguments which make clear that the addressee is already committed to implement the conclusion's plan. These strongest reasons can't *make* the addressee implement that plan, but they can show the addressee that he committed himself to implement it. Non-deductive directive illocutionary arguments will provide stronger or weaker reasons to implement the conclusion's plan, but they *invite* the addressee to make a commitment instead of showing that the commitment is already "in force."

It is instructive to compare the deductively correct argument above with the following directive illocutionary argument:

> Mark, you are morally obliged to close the windows if it rains while you are at home. It is raining now, and you are at home. So close the windows.

This most recent argument is not deductively correct, because obligation is not commitment. Being obliged to close the windows seems to me to be a good reason for closing them, but this doesn't mean that Mark must already be committed to do so.

Considering the different types of illocutionary acts and their associated arguments and derivations enables us to look for, and to recognize, parallels between the different types of acts and arguments. For example, we might wonder whether the distinction between strong and weak entailment that we find in directive locutionary arguments is also reflected in a distinction between strong and weak entailment relations linking statements, and whether any such distinction is important for assertive acts.

4   Commissive arguments

In performing a directive illocutionary act, a speaker attempts to get her addressee to implement a plan, after the addressee first commits himself to do this. In a (sincere) *commissive illocutionary act*, the speaker (writer, thinker) commits *herself* to implement a plan.

For commissive acts, there are also locutionary arguments, deductive derivations, and illocutionary arguments. The premises of a commissive locutionary argument can either be all plans or a mixture of plans and statements, and the premises of a commissive illocutionary argument can be assertive illocutionary acts (assertions and denials) or commissive illocutionary acts. Someone's *directive* illocutionary argument can make clear to an addressee that he is (already) committed to implement a plan, but such an argument cannot commit the addressee to do anything. With commissive illocutionary arguments, the premises can either show that the arguer is already committed to implement the conclusion's plan, or they can actually commit the arguer to implement that plan.

In this illocutionary argument:

> I have promised to close the windows if it rains while I am at home. It is raining, and I am at home. So I will close the windows now.

the premises show to the arguer that she has a prior commitment to close the windows, while in this argument:

> I will close the windows if it rains while I am at home. But it is raining, and I am at home. So I will close the windows now.

it is the commissive premiss together with the assertive premiss which give rise to the commitment to close the windows. Commissive illocutionary arguments, both deductive and non-deductive, are used to carry out *practical reasoning*.

In all three of the categories of illocutionary acts that we are considering, there are three associated types of argument/derivation. The locutionary arguments are valid or not, the deductive derivations are sound or not, and the illocutionary arguments are deductively correct or not. Deductive correctness depends in one way or another on rational commitment. Although I have been focusing on deductive arguments and derivations, there are three analogous types of non-deductive arguments and derivations. In actual practice, or "real-life" situations, only illocutionary arguments play important roles. For illocutionary arguments are the arguments that occur outside of logic books and logic classes. (However, the arguments that establish results like soundness and completeness for a deductive system are real, illocutionary, arguments.)

It is clear that locutionary acts are important primarily because of the illocutionary acts they are used to constitute, and that locutionary arguments and deductive or semantic derivations are important because of their relevance for illocutionary arguments. It is unfortunate that illocutionary arguments have received so little attention. But the conceptual framework that has been articulated here highlights some tasks that remain to be carried out, and provides some guidance as to how this can be done.

References

Kearns, J. T. (2006). Conditional Assertion, Denial, and Supposition as Illocutionary Acts. *Linguistics and Philosophy*, 29, 455–485.

Kearns, J. T. (2007). An Illocutionary Logical Explanation of the Liar Paradox. *History and Philosophy of Logic*, 28, 31–66.

Kearns, J. T. (2009a). An Illocutionary Conception of Syntax, Semantics, and Pragmatics. *Studies in Logic*, 2, 1–19.

Kearns, J. T. (2009b). Using Illocutionary Logic to Understand Vagueness. *Logique & Analyse*, *207*, 219–238.

Searle, J. R. (1969). *Speech Acts: an Essay in the Philosophy of Language*. Cambridge: Cambridge University Press.

Searle, J. R. (1985). *Expression and Meaning: Studies in the Theory of Speech Acts*. Cambridge: Cambridge University Press.

John T. Kearns
Department of Philosophy and Center for Cognitive Science, University at Buffalo, State University of New York
USA
E-mail: `kearns@buffalo.edu`

An Alternative Approach to Truth-value Semantics: *More or Less True than* and Pairwise Valuations

ROSSELLA MARRANO

Abstract: The truth-value semantics for propositional logics is given by a set of logical valuations, namely functions that map each propositional variable to one of the truth values and, in most cases, behave truth-functionally with respect to the connectives. In this paper we put forward, as an alternative approach, a semantics based on binary comparisons of sentences with respect to their truth. The key step consists in shifting the focus from pointwise valuations, which typically feature in standard truth-value semantics, to pairwise valuations based on comparative judgements. This paper provides grounds for (i) axiomatically defining pairwise valuations and pairwise semantics for a wide range of logics, and (ii) investigating the relation with the truth-value semantics.

Keywords: truth values, many-valuedness, truth ordering

1 Introduction

We are interested in expressing comparisons between sentences with respect to their truth, such as for example

"the sentence ϕ is *less (more) true than* the sentence ψ".

Comparative judgements of this kind represent an alternative method for evaluating the truth of the sentences of a formal language. The truth value of a sentence is determined by comparing it with other sentences in the language.

The standard approach consists rather in evaluating sentences by assigning them a certain value, called *truth value*. A semantics for a specific logic is then obtained by considering all the possible assignments of values to sentences. We will refer to this as the *truth-value semantics*. This approach dates back to Frege who introduced truth values as a special kind of objects representing possible denotations for sentences. According to Frege, there were just two of such objects: *the True* and *the False*. The functional approach turned out to be a powerful tool which easily allows for generalizations, starting with the cardinality of the set of truth values. Indeed the assumption that there are only two truth values can be, and has been, dropped. This leads to consider values other than true and false, ranging from three-valued logics to infinite-valued logics.

Many-valuedness poses new interesting philosophical challenges. One of the main issues is the status of the additional values, namely how to interpret them and whether it makes sense at all to call them *truth* values. Issue that becomes particularly pressing when infinitely many truth values, or degrees of truth, are considered (see Smith, 2008 for a discussion). The comparative perspective we propose in this paper provides a way for dealing with many-valuedness while avoiding explicit recourse to the set of truth values. In addition, the comparative perspective ultimately justifies the formulation in terms of truth values, because by using a measure-theoretical approach it can be showed that sets of comparative judgements can, under certain conditions, count as valuation functions.

The key step consists in taking as primitive a binary relation on the set of sentences interpreted as *more or less true than*. The first contribution of this paper is to lay down the formal conditions these sets of comparisons should meet in order to constitute a semantics for a specific logic, which we will refer to as *pairwise semantics*. The second contribution of this paper is to show that this semantics is compatible with the truth-value semantics to the extent that the axiomatic conditions defining the relation *more or less true than* are sufficient for it to be representable by a valuation function.

2 Notation and setup

Let $\mathcal{L}$ be a propositional language and $\mathcal{SL}$ the set of sentences built recursively by means of a binary connective $\to$ for implication, along with the constant $\bot$ for *falsum*. $(\mathcal{SL}, \to, \bot)$ is the algebra of terms. As usual negation, constant for *verum*, disjunction, conjunction and double implication are defined as $\neg \phi := \phi \to \bot$, $\top := \neg \bot$, $\phi \vee \psi := \neg \phi \to \psi$, $\phi \wedge \psi := \neg(\neg \phi \vee \neg \psi)$ and $\phi \leftrightarrow \psi := (\phi \to \psi) \wedge (\psi \to \phi)$, respectively.

Let $\mathbf{L} = \langle \mathcal{SL}, \vdash \rangle$, with $\vdash \subseteq \mathcal{SL}^2 \times \mathcal{SL}$, be an arbitrary logic over the language $\mathcal{SL}$ satisfying the following Assumptions.

1. $\mathbf{L}$ is a *Tarskian logic*, i.e. the deducibility relation $\vdash$ satisfies for every $\Gamma \subseteq \mathcal{SL}$ and every $\phi, \psi \in \mathcal{SL}$:

(REF) $\phi \in \Gamma$ implies $\Gamma \vdash \phi$,

(MON) $\Gamma \subseteq \Delta$ and $\Gamma \vdash \phi$ imply $\Delta \vdash \phi$,

(TRA) $\Gamma \vdash \phi$ and $\Gamma, \phi \vdash \psi$ imply $\Gamma \vdash \psi$.

(STR) for every endomorphism σ of $\mathcal{SL}$, $\Gamma \vdash \phi$ implies $\sigma(\Gamma) \vdash \sigma(\phi)$.

(FIN) $\Gamma \vdash \phi$ implies that there is a finite $\Gamma' \subseteq \Gamma$ such that $\Gamma' \vdash \phi$.

Also, **L** enjoys a deduction theorem in one of the following form

Theorem (Ordinary Deduction Theorem — DT) *If $\Gamma, \phi \vdash \psi$ then $\Gamma \vdash \phi \to \psi$.*

Theorem (Local Deduction Theorem — LDT) *If $\Gamma, \phi \vdash \psi$ then there is $n \in \omega$ such that $\Gamma \vdash \phi^n \to \psi$, where ϕ^n is an abbreviation for $\underbrace{\phi \wedge \cdots \wedge \phi}_{n \text{ times}}$.*

2. **L** is algebraizable in the sense of (Blok & Pigozzi, 1989), in particular strongly and regularly algebraizable (see Czelakowski, 2001, Definition 5.1.1, p. 352). This means that the class of algebra characterising the logic form a variety (strongly algebraizable) and that each algebra A has a distinguished element, $\top$, acting as designated (regularly algebraizable). We also assume $\top$ to be the interpretation of the logical symbol $\top$ of the algebra of terms. Let $\mathit{Var}_\vdash$ denote the variety characterising **L**.

3. All $A \in \mathit{Var}_\vdash$ are partially ordered as follows[1]

$$\forall a, b \in A \quad a \leq_V b :\Leftrightarrow a \to b = \top.$$

Moreover, $\leq_V$ is bounded, namely

$$\bot \to a = \top,$$
$$a \to \top = \top.$$

---

[1] For the sake of readability we denote the operations of the algebras in the variety characterising the logic **L** by using the same symbols of the connectives of $\mathcal{SL}$, for example $\to$ and $\bot$.

4. Let $\mathcal{SL}_0$ be the set of sentences in which just connectives and logical constants occur. For all $\Gamma \subseteq \mathcal{SL}_0$ and all $\phi \in \mathcal{SL}$

$$\Gamma \vdash \phi \Leftrightarrow \Gamma \vdash_{CL} \phi,$$

where $\vdash_{CL}$ is the deducibility relation of classical propositional logic. In other words, the logic of $\bot$ and $\top$ is classical.

Assumption 3. can also be expressed syntactically. Building on (Cintula, 2006), we could equivalently require

3'. **L** is a *weakly implicative logic*, namely the connective $\to$ satisfies

1. reflexivity: $\vdash \phi \to \phi$
2. transitivity: $\vdash \phi \to \psi, \psi \to \chi \vdash \phi \to \chi$
3. congruence: $\phi_1 \leftrightarrow \phi_2, \psi_1 \leftrightarrow \psi_2 \vdash (\phi_1 \to \psi_1) \leftrightarrow (\phi_2 \to \psi_2)$
4. MP: $\phi, \phi \to \psi \vdash \psi$.

Those conditions over the implication guarantee that for any algebra in the algebraic semantics of these logics one can define an order relation from the implication as in assumption 3. The fact that the order of truth values is bounded can be expressed syntactically as

5. $\vdash \bot \to \phi$,
6. $\vdash \phi \to \top$.

For the purposes of this paper, we restrict attention to logics which have as intended semantics a set of truth values which is partially ordered by the implication and with a greatest and smallest element, representing absolute truth and absolute falsity, respectively, which behave classically. This class of logics is broad enough to contain for example classical logic, finite-valued and infinite-valued logics, but also intuitionistic logic and substructural logics.

3 Pairwise L-valuations

To evaluate sentences *pairwise* means to perform comparative judgements of the form "ϕ is less (more) true than ψ". As anticipated, in order to formalise the idea of pairwise valuation we take as primitive a binary relation over the set of sentences, $\preceq \subseteq \mathcal{SL}^2$, interpreted as *no more true than*. As usual we define

$$\phi \prec \psi :\Leftrightarrow \phi \preceq \psi \text{ and not } \psi \preceq \phi,$$

and

$$\phi \sim \psi :\Leftrightarrow \phi \preceq \psi \text{ and } \psi \preceq \phi.$$

These relations are interpreted as *less true than* and *as true as*, respectively.

Pairwise valuations for a specific logic **L** satisfying the assumptions are defined in the following

Definition 1 *A relation* $\preceq \subset \mathcal{SL}^2$ *is a* pairwise **L**-valuation *if and only if*

(A.1) $\preceq$ *is transitive:* $\phi \preceq \chi, \chi \preceq \psi \Rightarrow \phi \preceq \psi$,

(A.2) $\sim$ *is a congruence:* $\phi_1 \sim \psi_2, \psi_1 \sim \psi_2 \Rightarrow \phi_1 \to \psi_1 \sim \phi_2 \to \psi_2$,

(A.3) $\vdash \phi \Rightarrow \phi \sim \top$,

(A.4) $(\phi \to \psi) \sim \top \Leftrightarrow \phi \preceq \psi$.

Pairwise valuations are comparisons, so we assume them to be transitive relations (A.1). Moreover, the derived relation *being as true as* is compatible with the underlying algebraic structure given by the connectives (A.2). This reflects the compositional nature of truth, namely the fact that truth values of complex sentences are determined by the truth values of the components. It is immediate to notice that $\sim$ is congruent also with respect to the defined sentential connectives. We also assume the relation to be *sound* with respect to the underlying logic, namely if a sentence is provable in the logical system (if it is a theorem) then it should be evaluated as maximally true (A.3). This axiom explicitly anchors the relation which expresses truth comparisons to the underlying logic. This may be considered an unwelcome feature for a semantic notion to possess, however at this level of generality it guarantees a sort of neutrality with respect to the chosen logic. When a specific logic is fixed, as far as the logic is finitely axiomatizable, then the axiom at stake can be removed in favour of a list of suitable conditions. It might be considered an 'abbreviation' for those conditions. In addition, we require the truth order to coincide with the order of truth values given by the implication. Axiom (A.4) also states the truth condition for the implication: an implicative sentence is true if and only if the truth value of the antecedent is less than or equal to the truth value of the consequent.

Immediate structural properties can be proved:

Proposition 1

1. $\preceq$ *is reflexive:* $\phi \preceq \phi$,

2. $\preceq$ *is bounded:* $\bot \preceq \phi \preceq \bot$.

The following propositions give us more details about the relation between pairwise **L**-valuations and the underlying logic.

Proposition 2 *If $\preceq$ is a pairwise **L**-valuation then the following hold*

1. $\vdash \phi \to \psi \Rightarrow \phi \preceq \psi$,

2. $\vdash \phi \leftrightarrow \psi \Rightarrow \phi \sim \psi$.

Proposition 3

1. $\phi \vdash \psi, \phi \sim \top \Rightarrow \psi \sim \top$,

2. $\Gamma \vdash \phi, \forall \gamma \in \Gamma \; \gamma \sim \top \Rightarrow \phi \sim \top$.

Proposition 2 establishes a crucial relation between the congruence *being as true as* and the relation of logical equivalence defined as

$$\phi \equiv \psi :\Leftrightarrow \vdash \phi \leftrightarrow \psi,$$

which is a congruence over $\mathcal{SL}$ by virtue of the assumptions. Proposition 3 states that a sort of strong soundness holds: if a sentence ϕ is a syntactical consequence of a set of sentences Γ then ϕ is absolutely true whenever each sentence in Γ is such according to the pairwise valuation.

4 Pairwise semantics

The truth-value semantics for a logic **L** is given by a set of logical valuations, namely truth-functional functions that map each propositional variable to one of the truth values.

Definition 2 *A pointwise **L**-valuation is a homomorphism* $h \colon \mathcal{SL} \to A$, *where* $A \in Var_\vdash$.

Let $\mathbb{V}$ be the set of all pointwise **L**-valuations containing for all $A \in Var_\vdash$ all the homomorphisms $h \colon \mathcal{SL} \to A$. Semantic consequence is defined as follows

Definition 3 *For all* $\Gamma \subseteq \mathcal{SL}$ *and* $\phi \in \mathcal{SL}$

$$\Gamma \models_V \phi :\Leftrightarrow \forall h \in \mathbb{V} \text{ if } \forall \gamma \in \Gamma \ h(\gamma) = \top \text{ then } h(\phi) = \top.$$

Semantic consequence is defined in terms of preservation of absolute truth ($\top$) under all the possible valuations. It is known that the relation thus defined is reflexive, monotone, transitive, and structural, that is $\models_V$ is a Tarskian consequence relation over $\mathcal{SL}$. Furthermore, being $\models_V$ defined in terms of the algebras in the variety characterising **L**, it is sound and complete with respect to the logic.

Proposition 4 *For all* $\Gamma \subseteq \mathcal{SL}$ *and* $\phi \in \mathcal{SL}$

$$\Gamma \models_V \phi \Leftrightarrow \Gamma \vdash \phi.$$

Since truth comparisons are by all means semantic valuations, we consider *families* of pairwise valuations, in the same way we consider all the possible assignments in the standard functional setting. Let $\mathcal{P}$ be the set of all the possible pairwise **L**-valuations. For any family of pairwise valuations $\mathcal{F} \subseteq \mathcal{P}$ we can consider the intersection over $\mathcal{F}$, namely

$$\bigcap \mathcal{F} = \{(\phi, \psi) \subseteq \mathcal{SL}^2 | \text{ for all } \preceq \text{ in } \mathcal{F} \ \phi \preceq \psi\},$$

and the corresponding preorder

$$\phi \preceq_\mathcal{F} \psi :\Leftrightarrow (\phi, \psi) \in \bigcap \mathcal{F}.$$

It is easy to verify that $\preceq_\mathcal{F}$ is a pairwise **L**-valuation.

The *intersection* over a family of preorders plays an important role. The pairwise valuation resulting from the intersection intuitively represents the set of comparisons which hold under all the possible interpretations, or in all possible worlds. In the standard truth-value semantics for propositional logic, the sentences which are true under all the possible assignments are called *tautologies* or *logical truths*. Those have the property of remaining true under all uniform substitutions. Substitution invariance corresponds to the property of being analytical. From a comparative perspective, we deal with families of possible pairwise valuations. Analogously, the preorder resulting from the intersection of all pairwise valuation expresses the comparative judgements which hold *analytically* or *logically*. For the reasons just stated, substitution invariance is a necessary requirement and we take it as a criterion of *admissibility* of a family of pairwise valuation.

Definition 4 *A family of pairwise valuations* $\mathcal{A} \subseteq \mathcal{P}$ *is* admissible *if and only if* $\preceq_\mathcal{A}$ *is substitution-invariant, i.e. for all endomorphisms* $\sigma \colon \mathcal{SL} \to \mathcal{SL}$

$$\phi \preceq_\mathcal{A} \psi \Rightarrow \sigma(\phi) \preceq_\mathcal{A} \sigma(\psi).$$

An admissible family of pairwise valuations induces in a natural way a semantics for the logic at stake. The notions of tautology and semantic consequence can be defined as follows:

Definition 5

$$\Gamma \models_\preceq \phi :\Leftrightarrow \text{ for all pairwise } \mathbf{L}\text{-valuations } \preceq \in \mathcal{A}, \text{ if } \forall \gamma \in \Gamma \; \gamma \sim \top$$
$$\text{then } \phi \sim \top$$
$$\Leftrightarrow \text{ if } \forall \gamma \in \Gamma \; \gamma \sim_\mathcal{A} \top \text{ then } \phi \sim_\mathcal{A} \top.$$
$$\models_\preceq \phi :\Leftrightarrow \text{ for all pairwise } \mathbf{L}\text{-valuations } \preceq \in \mathcal{A} \; \phi \sim \top$$
$$\Leftrightarrow \phi \sim_\mathcal{A} \top.$$

Those are classical definitions in which validity is accounted for in terms of absolute truth preservation: whenever all the premises are absolutely true, the conclusion should be absolutely true as well. The semantics thus defined can be proved to be strongly sound and complete with respect to the logic **L**. Soundness follows directly from proposition 3 and the definition of $\models_\preceq$. In what follows we prove strong completeness.

Theorem 1 (Strong completeness) *For all* $\Gamma \subseteq \mathcal{SL}$ *and* $\phi \in \mathcal{SL}$

$$\Gamma \models_\preceq \phi \Rightarrow \Gamma \vdash \phi.$$

Proof. We prove that if $\Gamma \nvdash \phi$ then there exists an admissible family and a pairwise **L**-valuation such that $\forall \gamma \in \Gamma \; \gamma \sim \top$ and $\phi \nsim \top$. Let

$$\phi \preceq \psi :\Leftrightarrow \Gamma \vdash \phi \to \psi.$$

We can verify that

(i) $\preceq$ is a pairwise **L**-valuation.

(A.1),(A.2) $\preceq$ is a congruent preorder because $\to$ is reflexive, transitive, and congruent (see point 3' of the assumptions).

(A.3) $\vdash \phi \Rightarrow \vdash \top \leftrightarrow \phi \Rightarrow \Gamma \vdash \top \leftrightarrow \phi \Rightarrow \phi \sim \top.$

(A.4) $\phi \to \psi \sim \top \Leftrightarrow \Gamma \vdash (\phi \to \psi) \leftrightarrow \top \Leftrightarrow \Gamma \vdash \phi \to \psi \Leftrightarrow \phi \preceq \psi$.

(ii) $\Gamma \not\preceq_{\preceq} \phi$, namely $\forall \gamma \in \Gamma\ \gamma \sim \top$ and $\phi \not\sim \top$. Notice that $\Gamma \vdash \phi \Leftrightarrow \Gamma \vdash \top \leftrightarrow \phi \Leftrightarrow \top \sim \phi$, that is to say the defined pairwise valuation gives value $\top$ to all and only the sentences derivable from Γ.

(iii) There exists an admissible family $\mathcal{F}$ such that $\preceq$ is in $\mathcal{F}$. For all $\Gamma \subseteq \mathcal{SL}$ define $\preceq_\Gamma$ such that

$$\phi \preceq_\Gamma \psi :\Leftrightarrow \Gamma \vdash \phi \to \psi.$$

Consider the family $\mathcal{F} \subseteq \mathcal{P}$ of pairwise L-valuations thus defined. We prove that $\mathcal{F}$ is admissible. Let $\preceq_\mathcal{F}$ be the intersection preorder over $\mathcal{F}$. Notice that $\phi \preceq_\mathcal{F} \psi \Leftrightarrow$ for all $\Gamma \subseteq \mathcal{SL}\ \Gamma \vdash \phi \to \psi$. Furthermore, for all $\Gamma \subseteq \mathcal{SL}\ \Gamma \vdash \phi \to \psi$ implies and is implied by $\vdash \phi \to \psi$. And $\vdash$ is substitution-invariant because it is equivalent to $\models_V$, which satisfies substitution invariance. Then, we have that also $\preceq_\mathcal{F}$ is such.

□

Soundness and completeness results guarantee that pairwise valuations supply **L** with an adequate semantics, alternative to the standard truth-value semantics, though still defined in terms of (absolute) truth preservation. The fact that this semantics is sound comes as no surprise given that by assumption each pairwise valuation evaluates theorems of the logic as absolutely true (axiom (A.3)). We are not assuming the right-to-left direction of axiom (A.3) so that also sentences which are not tautology can be considered absolutely true under a certain valuation. However, a form of completeness hold: if a sentence is absolutely true under all the possible pairwise valuations in an admissible family then it is also provable in the logic. Also, strong completeness holds, namely completeness with respect to a set of premises: whenever a sentence is a semantic consequence of a set of sentences, there is a proof of this sentence in the system.

The proof of completeness bring out the centrality of theories and the possibility of defining pairwise valuations in terms of the implicative relations holding in a set of sentences. This can be made explicit by noticing that there is a correspondence between the pairwise L-valuations as defined in definition 1 and the deductively closed sets or theories of the logic **L**. This formal relation do not undermine the interest of investigating truth rankings as primitive objects. On the contrary, it adds depth and weight to the analysis. On the one hand, formal interactions with other logical notions constitute a testing ground for the definitions. On the other hand, those structural similarities confirm that we are bringing to the foreground notions which are deeply interwoven with the logical structure of our systems. Re-elaborating them in a form that allows a philosophical interpretation in terms of the relation *being more true than* is one of the contributions of this investigation.

5 Representation theorem

We are interested in proving that any pairwise **L**-valuation in a given admissible family can be represented by a pointwise **L**-valuation. However before that, it is of interest pointing out that, since we are assuming that a partial order is defined over the set of truth values, for any pointwise **L**-valuation there exists a corresponding pairwise **L**-valuation. In other words, each logical valuation of the logic **L** can be expressed by means of a set of comparisons between sentences.

Theorem 2 (From pointwise to pairwise) *Each pointwise **L**-valuation induces a pairwise **L**-valuation. The family $\mathbb{V}$ of pointwise **L**-valuations induces an admissible family of pairwise **L**-valuations.*

Proof. For all h in $\mathbb{V}$ define

$$\phi \preceq_h \psi :\Leftrightarrow h(\phi) \leq_V h(\psi).$$

We prove that

(i) $\preceq_h$ is a pairwise **L**-valuation.

(A.1),(A.2) $\preceq_h$ is a congruent preorder because $\leq_V$ is such.

(A.3) $\vdash \phi \Rightarrow \models_V \phi \Rightarrow h(\phi) =_V \top \Rightarrow \phi \sim_h \top$.

(A.4) $(\phi \to \psi) \sim_h \top \Leftrightarrow h(\phi \to \psi) =_V \top \Leftrightarrow h(\phi) \leq_V h(\psi) \Leftrightarrow \phi \preceq_h \psi$.

(ii) The set $\mathcal{F}$ of pairwise **L**-valuations generated in this way is admissible, namely the intersection preorder $\bigcap \mathcal{F}$ is substitution-invariant. This rests on the fact that $\models_V$ is such.

□

More or Less True than and Pairwise Valuations

This is to some extent the easy direction of the representation result. The set of truth values, irrespectively of its cardinality, usually comes with an ordering structure, called *truth ordering*. For instance, in the classical case, truth values constitute a lattice in which *false* is less true than *true*; or in the infinite-valued case degrees of truth form a bounded chain. In general, we take this order to be the natural order of the variety characterising the logic, determined from the implication. This natural order induces a truth ranking over the set of sentences, as theorem 2 states. Notice that the induced pairwise valuation is more than a preorder: it is a partial order. In some cases, depending on the truth-value semantics, it can be also linear or Archimedean. This gives the chance to make an important point in favour of the qualitative perspective. Defining the truth ranking directly over the set of sentences, instead of considering the ranking endowed in the set of truth values, allows us to take some of the mathematical structure off and to get rid of non-essential mathematical properties. This might be a move rich in philosophical and methodological relevance.

The main focus is still to show that it is possible to move from pairwise to pointwise valuations, that is to say to show that if a set of comparative judgements satisfies certain requirements, then it can be represented by a function assigning truth values which are ordered in a compatible way. More precisely, we want the natural order of the set of truth values, namely the order of the variety characterising the logic, to preserve the original pairwise valuation:

Definition 6 *A pointwise* **L**-*valuation* $h\colon \mathcal{SL} \to A$ *represents a pairwise* **L**-*valuation* $\preceq\, \subseteq \mathcal{SL}^2$ *if and only if for all* $\phi, \psi \in \mathcal{SL}$

$$\phi \preceq \psi \Rightarrow h(\phi) \leq_V h(\psi).$$

Theorem 3 (From pairwise to pointwise) *Given an admissible family of pairwise* **L**-*valuations* $\mathcal{A} \subseteq \mathcal{P}$, *for every* $\preceq$ *in* $\mathcal{A}$ *there exists at least one pointwise* **L**-*valuation representing it.*

Proof. The proof rests on a central algebraic result (see Burris & Sankappanavar, 2000, theorem 8.6, p. 64.):

Lemma 1 (Birkhoff's subdirect representation theorem) *Every algebra* **A** *is isomorphic to a subdirect product of subdirectly irreducible algebras (which are homomorphic images of* **A***).*

Accordingly, given a pairwise valuation $\preceq$ in an admissible family, the key-step is to construct an algebra on $(\mathcal{SL}, \preceq)$. In particular, we consider the quotient algebra *modulo* $\sim$, obtained by taking as universe the quotient set $\mathcal{SL}/\sim = \{[\phi]_\sim | \phi \in \mathcal{SL}\}$, where $[\phi]_\sim = \{\psi \in \mathcal{SL} | \psi \sim \phi\}$. The operations

$$[\phi]_\sim \stackrel{\sim}{\to} [\psi]_\sim := [\phi \to \psi]_\sim$$

$$\stackrel{\sim}{\bot} := [\bot]_\sim := \bot$$

are well defined because $\sim$ is a congruence over $\mathcal{SL}$.

The fact that the truth order refines the order given by the logical equivalence relation, i.e. the fact that

$$\phi \equiv \psi \Rightarrow \phi \sim \psi,$$

allows us to relate the quotient algebra *modulo* $\sim$ with the algebra $(\mathcal{SL}/\equiv, \to, 0)$, known as the Lindenbaum algebra for **L**. This algebra exists by virtue of the assumptions over **L** and, also, we know that it is in *Var*$_\vdash$. Let $q_\equiv$ and $q_\sim$ be the canonical maps from $\mathcal{SL}$ to the quotients $\mathcal{SL}/\equiv$ and $\mathcal{SL}/\sim$, respectively. Notice that these functions are onto. This being in place, the relation between the structures is sketched in the following diagram:

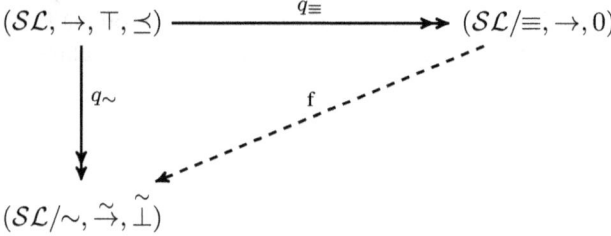

In order to make the diagram commute, we define a function $f \colon \mathcal{SL}/\equiv \to \mathcal{SL}/\sim$ as follows

$$\forall \phi \in \mathcal{SL} \quad f([\phi]_\equiv) = [\phi]_\sim.$$

It is straightforward to verify that:

Proposition 5

1. f is well defined,

2. f is onto,

3. $f(q_\equiv(\phi)) = q_\sim(\phi)$.

4. $f: \mathcal{SL}/\!\!\equiv\; \to \mathcal{SL}/\!\!\sim$ *is a homomorphism, namely*

   (a) $f([\phi]_\equiv \to [\psi]_\equiv) = f([\phi]_\equiv) \stackrel{\sim}{\to} f([\psi]_\equiv)$,

   (b) $f(0) = \widetilde{\bot}$.

Since varieties of algebras are closed under homomorphic images, the following holds:

Lemma 2 $(\mathcal{SL}/\!\!\sim, \stackrel{\sim}{\to}, \widetilde{\top}) \in \mathit{Var}_\vdash$.

By applying lemma 1 we know that there exists a family I of congruences over $\mathcal{SL}/\!\!\sim$ together with a homomorphism $e: \mathcal{SL}/\!\!\sim\; \to \prod_{\theta \in I} A_\theta$, where A_θ is $\mathcal{SL}/\!\!\sim/\theta$ with θ in I. The homomorphism e is such that the composition with each projection function is onto, as pictured in figure 1, and such that each A_θ is subdirectly irreducible. In addition, each $A_\theta \in \mathit{Var}_\vdash$, because for lemma 2 we know that $(\mathcal{SL}/\!\!\sim, \stackrel{\sim}{\to}, \widetilde{\bot}) \in \mathit{Var}_\vdash$ and variety are closed under quotients. Also, $\leq_V$, the natural order of A_θ, preserves $\preceq$. Since $(\mathcal{SL}/\!\!\sim, \stackrel{\sim}{\to}, \widetilde{\top}) \in \mathit{Var}_\vdash$ then it has its natural partial order defined as follows

$$[\phi]_\sim \leq_V [\psi]_\sim :\Leftrightarrow [\phi]_\sim \stackrel{\sim}{\to} [\psi]_\sim = [\top]_\sim.$$

Moreover, $\preceq$ induces in a natural way partial order $\leq_\sim$ on the quotient set defined as

$$[\phi]_\sim \leq_\sim [\psi]_\sim :\Leftrightarrow \exists \phi_i \in [\phi]_\sim, \psi_i \in [\psi]_\sim \text{ such that } \phi_i \preceq \psi_i,$$

which by definition preserves $\preceq$. However, axiom (A.4) guarantees that the two orderings coincide. Homomorphisms preserve the natural orders of the algebra, so the claim follows. The function $v_\preceq$ obtained by composing $q_\sim$ and h is the desired pointwise **L**-valuation representing $\preceq$.

□

Theorem 3 shows that the set of conditions over pairwise valuations is sufficient for guaranteeing the existence of a matching pointwise valuation, namely a logical valuation for that logic evaluating sentences in the intended truth-value semantic which preserves the original truth ranking.

Figure 1: From **L**-pairwise to **L**-pointwise valuation

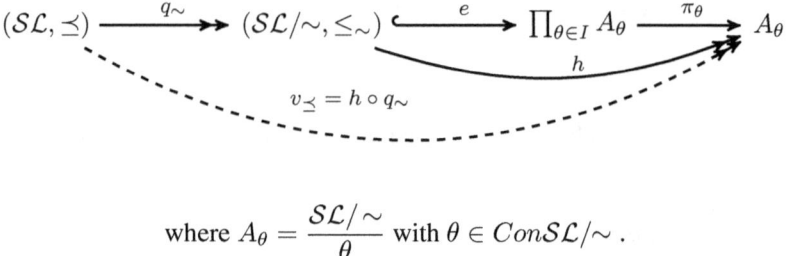

where $A_\theta = \dfrac{\mathcal{SL}/\sim}{\theta}$ with $\theta \in Con\mathcal{SL}/\sim$.

6 Conclusion

Pointwise and pairwise valuations are alternative methods for evaluating the truth of sentences of a formal system. The former evaluate sentences by assigning them a truth value, whereas the latter by means of binary comparisons. This paper is concerned with the relationships between those two notions, the central problem being to establish representation results, i.e. to isolate a set of conditions over pairwise valuations sufficient for guaranteeing the existence of pointwise valuations representing them.

The idea of pairwise valuations has been in the first place called for in terms of the intuitive and preformal notion of *being less (more) true than* and then formalised in definition 1. We showed that the semantics induced by pairwise valuations is strongly sound and complete with respect to the logic (theorem 1). Moreover, we proved that the pairwise semantics yields the standard intended semantics, to the extent that the axiomatic conditions defining the relation *more or less true than* are sufficient for it to be representable by a valuation function of that logic (theorem 3).

The desired pointwise valuation is obtained by composition of homomorphisms. The canonical map $q_\sim$ is already a representing pointwise valuation. However, we have seen that something more can be done, i.e. we can construct pointwise valuations evaluating sentences in more fundamental algebras, the irreducible ones, which are obtained by refining the quotient, namely by composing the congruence $\sim$ with other congruences θ. Irreducible algebras in the variety characterising the logic play the role of the algebra of truth values: their universe is the set of truth values and their operations (which are of the same signature as sentential connectives) are truth functions assigning truth values to complex sentences on the basis of the truth values of their components. This set of truth values is partially ordered with an order induced by the implication. Furthermore, the fact that the final pointwise valuation is onto amounts to say that all the truth values in the algebra are actually needed in order to represent the truth ranking as a functional assignment.

Representation results like theorem 3 assure that if the sentences can be compared 'well enough' with respect to their truth, where 'well enough' is given by a set of definitory axiomatic conditions, then it is as if we attach them a specific truth value. The existence of a representation allows us to gain the mathematical convenience typical of the functional approach, in which the idea of being true to different extents in modelled by using special, usually numerical, objects, namely the truth values. Furthermore, there is a gain in philosophical plausibility in the idea of truth from comparison. It provides a viable way for dealing with logical valuedness without resorting to a set of objects whose nature and philosophical status might be questionable.

References

Blok, W., & Pigozzi, D. (1989). Algebraizable Logics. *Memoirs of the American Mathematical Society, 77*(396).

Burris, S., & Sankappanavar, H. (2000). *A Course in Universal Algebra*. Berlin: Springer.

Cintula, P. (2006). Weakly Implicative (Fuzzy) Logics I: Basic Properties. *Archive for Mathematical Logic, 45*, 673–704.

Czelakowski, J. (2001). *Protoalgebraic Logics*. Berlin: Kluwer Academic Publishers.

Smith, N. (2008). *Vagueness and Degrees of Truth*. Oxford: Oxford University Press.

Rossella Marrano
Scuola Normale Superiore
Italy
E-mail: rossella.marrano@sns.it

Classical Logic Through the Looking-glass

PETER MILNE

Abstract: In Lewis Carroll's *Through the Looking Glass and What Alice Found There*, Alice enters through a mirror into the realm reflected. It is, of course, left-right reversed but this is only the start of the fun and games when Alice explores the world on the other side of the mirror. Borrowing, if only in part, Carroll's theme of inversion, my aim is to take a look at classical logic in something of an inverted way, or, to be more exact, in three somewhat inverted ways. Firstly, I come at proof of the completeness of classical logic in the Lindenbaum-Henkin style backwards: I take for granted the existence of a set Σ for which it holds, for some formula ϕ, that $\psi \notin \Sigma$ if, and only if, $\Sigma \cup \{\psi\} \vdash \phi$ then read off the rules of inference governing connectives and quantifiers that most directly yield the desired (classical) semantic properties. We thus obtain *general elimination rules* and what I have elsewhere called *general introduction rules*. Secondly, the same approach lets us read off a different set of rules: those of the cut-free sequent calculus $\mathscr{S}'$ of (Smullyan, 1968). Smullyan uses this calculus in proving the Craig-Lyndon interpolation theorem for first-order logic (without identity and function symbols). By attending very carefully to the steps in Smullyan's proof, we obtain a strengthening: if $\phi \vdash \psi, \nvdash \neg\phi$ and $\nvdash \psi$ then there is an interpolant χ, a formula employing only the non-logical vocabulary common to ϕ and ψ, such that ϕ entails χ in the first-order version of Kleene's 3-valued logic and χ entails ψ in the first-order version of Graham Priest's Logic of Paradox. The result, which is hidden from view in natural deduction formulations of classical logic, extends, I believe, to first-order logic with identity. Thirdly, we look at a contraction-free "approximation" to classical propositional logic. Adding the general introduction rules for negation or the conditional leads to Contraction being a derived rule, apparently blurring the distinction between structural and operational rules.

Keywords: completeness proof for classical first-order logic, Lindenbaum-Henkin construction, general elimination rules, general introduction rules, Craig interpolation lemma, Kleene's strong three-valued logic, logic of paradox, Łukasiewicz's infinite-valued logic, contraction

Peter Milne

1 Introduction

Speaking of A. J. Ayer's criticism of Sartre's philosophy of *le néant*, the Norwegian philosopher Arne Næss said,

> Characteristically the critic's appeal in this case is not to our scientific sensibilities and the logical calculus of predicates but to that much respected tribunal in British philosophy, *Alice through the Looking Glass*. (Næss, 1968, p. 318)

To give it its proper title, this work, the second of the Reverend Charles Lutwidge Dodgson's Alice books, is *Through the Looking-Glass and What Alice Found There*. Alice enters through a mirror into the realm reflected. It is, of course, left-right reversed but this is only the start of the fun and games when Alice explores the world on the other side of the mirror.

I shall not invoke *Through the Looking-Glass* as a tribunal (however, exactly, one might do that). I want only to borrow, and only in part, Lewis Carroll's theme of inversion.[1] My aim is to take a look at classical logic in something of an inverted, back-to-front way, or, to be more exact, in three, distinct but related, somewhat back-to-front ways.

First I'll come at the Lindenbaum-Henkin proof of completeness backwards, obtaining rules with a view to showing that the maximal, consistent extension has the properties required of the set of formulas. Then I'll look at a refinement of the Craig-Lyndon Interpolation Theorem for Classical First-order Logic using a formulation of first-order logic derived from the first part. Thirdly, I'll develop what can reasonably be called a contraction-free variant of Classical First-order Logic. Replacing standard rules for negation or the conditional with the rules obtained in the first investigation, *i.e.*, by changing what would normally be thought of as operational rules, we re-introduce contraction (at the cost of proofs without the subformula property).

2 Ideal rules for proving completeness[2]

In outline the orthodox procedure when proving the completeness of classical logic in the Lindenbaum-Henkin style is as follows:

---

[1] On the looking-glass theme and inversion in Carroll's writings, see (Carroll, 1970, n. 4, pp. 180–83).

[2] This section is in part a reworking of material drawn from (Milne, 2008, 2010, 2015).

Classical Logic Through the Looking-glass

1. Starting from a (possibly empty) set of premises, Σ, and a formula, χ, that is *not* derivable from the set and given an enumeration of formulas in the language, one expands the premise set successively adding just those formulas whose addition does not lead to the derivability of the initially underivable sentence.

2. In the limit this procedure yields a set of sentences Σ_∞ for which, for all formulas ϕ of the language,

$$\phi \notin \Sigma_\infty \text{ if, and only if, } \Sigma_\infty \cup \{\phi\} \vdash \chi.$$

3. We then use the rules of the logic to show that Σ_∞ has exactly the closure properties it would have were it the set of formulas true (satisfied) in a model. As $\psi \notin \Sigma_\infty$, there is, then, a model in which all members of Σ are true and χ is not.

Let us go at this from the opposite end—Carrollian inversion! Suppose that we have a set of formulas possessing the closure properties of a set of formulas true (satisfied) in a model as classically understood. What would be the best rules for showing this?

You may think that this isn't a very precise question so let me show you what I have in mind. Consider conjunction, first of all. Here we need that $\phi \wedge \psi \in \Sigma_\infty$ if, and only if, $\phi \in \Sigma_\infty$ and $\psi \in \Sigma_\infty$. First, I'll contrapose this: $\phi \wedge \psi \notin \Sigma_\infty$ if, and only if, $\phi \notin \Sigma_\infty$ or $\psi \notin \Sigma_\infty$. Next, I'll recast this in terms of the Lindenbaum-Henkin condition of non-membership:

- if $\Sigma_\infty \cup \{\phi\} \vdash \chi$ then $\Sigma_\infty \cup \{\phi \wedge \psi\} \vdash \chi$;

- if $\Sigma_\infty \cup \{\psi\} \vdash \chi$ then $\Sigma_\infty \cup \{\phi \wedge \psi\} \vdash \chi$;

- if $\Sigma_\infty \cup \{\phi \wedge \psi\} \vdash \chi$ then $\Sigma_\infty \cup \{\phi, \psi\} \vdash \chi$.

What we are looking for are generally applicable rules of inference that guarantee this as directly as possible. What we read off the first condition is that when ϕ together with side premises entails χ then $\phi \wedge \psi$ together with those same side premises also entails χ. Likewise, from the second condition we read off that when ψ together with side premises entails χ then $\phi \wedge \psi$ together with those same side premises also entails χ. We set these out as follows:

Peter Milne

$$\frac{\phi \wedge \psi \quad \overset{[\phi]^m}{\underset{\chi}{\vdots}}}{\chi} m \wedge\text{-elimination (l);} \qquad \frac{\phi \wedge \psi \quad \overset{[\psi]^m}{\underset{\chi}{\vdots}}}{\chi} m \wedge\text{-elimination (r).}$$

These are no more than transcriptions of the standard elimination rules in what is sometimes called the *general elimination* format. (See Milne, 2015, p. 192 for references.) Terminology: we say that the conjunction occurs *categorically* and that the conjuncts occur *hypothetically* in these rules.

We read off the third condition that when $\phi \wedge \psi$ together with side premises entails χ then those same side premises together with ϕ and ψ suffice to entail χ. We may write this thus:

$$\frac{\phi \quad \psi \quad \overset{[\phi \wedge \psi]^m}{\underset{\chi}{\vdots}}}{\chi} m \wedge\text{-introduction.}$$

(Here the conjunction occurs hypothetically, the conjuncts categorically.)

From this, which promises to be the most straightforward case, we see that introduction rules are very different in form from the norm. Usually, the introduced connective occurs as main connective in a formula which stands as conclusion of the application of the rule:

$$\frac{\phi \quad \psi}{\phi \wedge \psi}.$$

In our rule the introduced connective occurs as main connective in a formula which is an assumption apt for discharge in the application of the rule. In effect, the standard rule is the special case when χ is $\phi \wedge \psi$.

In a small step in the direction of familiarity, disjunction gives us these conditions:

- if $\Sigma_\infty \cup \{\phi\} \vdash \chi$ and $\Sigma_\infty \cup \{\psi\} \vdash \chi$ then $\Sigma_\infty \cup \{\phi \vee \psi\} \vdash \chi$;
- if $\Sigma_\infty \cup \{\phi \vee \psi\} \vdash \chi$ then $\Sigma_\infty \cup \{\phi\} \vdash \chi$;
- if $\Sigma_\infty \cup \{\phi \vee \psi\} \vdash \chi$ then $\Sigma_\infty \cup \{\psi\} \vdash \chi$;

Classical Logic Through the Looking-glass

and hence these rules:

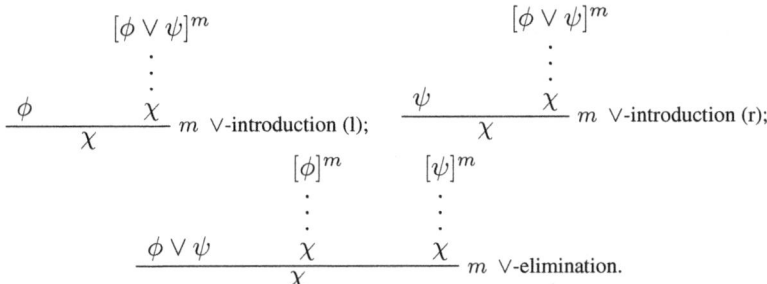

The standard ∨-elimination rule is already in general elimination format.

In the case of the conditional, we get something really different. First, we have that $\phi \to \psi \in \Sigma_\infty$ if, and only if, $\phi \notin \Sigma_\infty$ or $\psi \in \Sigma_\infty$. Contraposing, $\phi \to \psi \notin \Sigma_\infty$ if, and only if, $\phi \in \Sigma_\infty$ and $\psi \notin \Sigma_\infty$. Hence, recasting in terms of the condition of non-membership:

○ if $\Sigma_\infty \cup \{\psi\} \vdash \chi$ then $\Sigma_\infty \cup \{\phi, \phi \to \psi\} \vdash \chi$;

○ if $\Sigma_\infty \cup \{\phi \to \psi\} \vdash \chi$ then $\phi \in \Sigma_\infty$;

○ if $\Sigma_\infty \cup \{\phi \to \psi\} \vdash \chi$ then $\Sigma_\infty \cup \{\psi\} \vdash \chi$.

The first clause gives us the general elimination form of *modus (ponendo) ponens*:

$$\dfrac{\phi \to \psi \quad \phi \quad \begin{array}{c}[\psi]^m\\ \vdots\\ \chi\end{array}}{\chi}\ m \to\text{-elimination.}$$

The third clause too is straightforward in its import:

$$\dfrac{\psi \quad \begin{array}{c}[\phi \to \psi]^m\\ \vdots\\ \chi\end{array}}{\chi}\ m \to\text{-introduction (c).}$$

It's the second clause, and, in particular, what to do with that $\phi \in \Sigma_\infty$ in the consequent, that at first sight poses a problem. But we can think of it this way: the clause as a whole tells us that $\Sigma_\infty \cup \{\phi \to \psi\}$'s entailing χ

suffices for ϕ's belonging to Σ_∞. And ϕ belongs to Σ_∞ if, and only if, the assumption that adding ϕ to Σ_∞ lets us derive χ is equivalent to saying that Σ_∞ itself entails χ. Putting this in the form of a rule, we get:

$$\begin{array}{cc} [\phi]^m & [\phi \to \psi]^m \\ \vdots & \vdots \\ \chi & \chi \end{array}$$
$$\overline{\chi} \quad m \;\to\text{-introduction (a).}^3$$

Here's another way to think about this. If $\Sigma_\infty \cup \{\phi \to \psi\} \vdash \chi$ only if $\phi \in \Sigma_\infty$ then there's an incoherence in having both $\Sigma_\infty \cup \{\phi \to \psi\} \vdash \chi$ and $\Sigma_\infty \cup \{\phi\} \vdash \chi$, i.e., $\phi \notin \Sigma_\infty$. Now, *in context*, '$\Sigma_\infty \vdash \chi$' is a way of expressing that incoherence, for our starting point is, exactly, that $\Sigma_\infty \nvdash \chi$.

These both help when we turn to negation, as we now do. We have that $\neg\phi \in \Sigma_\infty$ if, and only if, $\phi \notin \Sigma_\infty$. We have, on the one hand, that that $\neg\phi \in \Sigma_\infty$ and that $\phi \in \Sigma_\infty$ are jointly incoherent. On the first way, this gives us the familiar $\neg$-elimination rule, *ex falso quodlibet*:

$$\frac{\phi \quad \neg\phi}{\chi} \quad \neg\text{-elimination.}$$

We have, on the other hand, that that $\neg\phi \notin \Sigma_\infty$ and that $\phi \notin \Sigma_\infty$ are jointly incoherent. On the second way, this gives us the Rule of Dilemma as $\neg$-introduction rule:

$$\begin{array}{cc} [\phi]^m & [\neg\phi]^m \\ \vdots & \vdots \\ \chi & \chi \end{array}$$
$$\overline{\chi} \quad m \;\neg\text{-introduction.}$$

Before we turn our attention to rules for the quantifiers, let's look at these propositional logic rules a little more closely. We've read the rules off the closure properties of a set of formulas true (satisfied) in a model as classically understood. If the rules capture those properties, rather than just being in some way consequences of them, then, in another backwards journey, we should be able to read the semantic constraints off the rules. And so we can. We read them as follows: label categorically occurring

---

[3]Elsewhere I have called this 'Tarski's Rule' (Milne, 2008, 2010) for it bears the same relation to the tautology sometimes called 'Tarski's Law' as the better known Peirce's Rule does to Peirce's Law.

subformulas as *true*, hypothetically as *false*; label formulas in which the connective of interest occurs the other way around—hypothetical = *true*, categorical = *false*. Doing this we revisit the utterly familiar:

- the ∧-elimination (l) rule tells us that $\phi \wedge \psi$ is false when ϕ is false;
- the ∧-elimination (r) rule tells us that $\phi \wedge \psi$ is false when ψ is false;
- the ∧-introduction rule tells us that $\phi \wedge \psi$ is true when ϕ and ψ are both true;
- the ∨-elimination rule tells us that $\phi \vee \psi$ is false when ϕ and ψ are both false;
- the ∨-introduction (l) rule tells us that $\phi \vee \psi$ is true when ϕ is true;
- the ∨-introduction (r) rule tells us that $\phi \vee \psi$ is true when ψ is true;
- the →-elimination rule tells us that $\phi \to \psi$ is false when ϕ is true and ψ is false;
- the →-introduction (a) rule tells us that $\phi \to \psi$ is true when ϕ is false;
- the →-introduction (c) rule tells us that $\phi \to \psi$ is true when ψ is true;
- the ¬-elimination rule tells us that $\neg \phi$ is false when ϕ is true;
- the ¬-introduction rule tells us that $\neg \phi$ is true when ϕ is false.

Now consider the binary connective with the truth-table (exclusive disjunction) in table 1.

$\phi + \psi$	ψ	
	t	f
ϕ t	f	t
ϕ f	t	f

Table 1: Truth-table for exclusive disjunction

We can read off two +-introduction rules. In the order top right, bottom left we get:

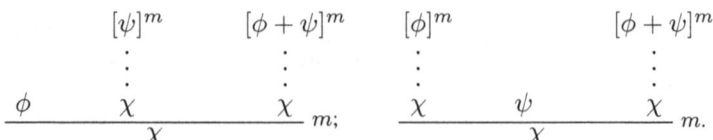

And we can read off two +-elimination rules. In the order top left, bottom right we get:

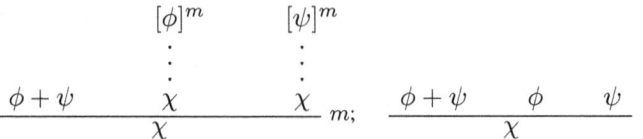

We can go backward and forward between introduction and elimination rules and truth-tables. (And we can do this for connectives of any *arity*; see (Milne, 2015, § 8.4.3) for more on this.)

We turn to the quantifiers and return to classical logic. We have that $\forall x \phi(x) \in \Sigma_\infty$ only if, for all terms t, $\phi(t) \in \Sigma_\infty$. Contraposing, $\forall x \phi(x) \notin \Sigma_\infty$ if, for some t, $\phi(t) \notin \Sigma_\infty$; and so, for any t, $\Sigma_\infty \cup \{\forall x \phi(x)\} \vdash \chi$ if $\Sigma_\infty \cup \{\phi(t)\} \vdash \chi$ which gives us the familiar elimination rule (recast in general elimination form):

$$\frac{\forall x \phi(x) \qquad \begin{array}{c} [\phi(t)]^m \\ \vdots \\ \chi \end{array}}{\chi} \; m \; \forall\text{-elimination.}$$

For the introduction rule for the universal quantifier—and the elimination rule for the existential quantifier—we *have* to take a more adventurous, and possibly less convincing, line; this is because parametric occurrences of names are not like ordinary individual constants—they really only exist in proofs, not in the model theory.[4] And tied in with that is the fact that we haven't really made provision for $\phi(a)$ occurring in the enumeration of formulas used when obtaining Σ_∞. Still, let's press on. (*Allez en avant, et la foi vous viendra!*) Somewhat in the style of Kit Fine's theory of arbitrary objects (Fine, 1985), we introduce a name 'a' which behaves syntactically as an individual constant but which we treat as though it were the name of a "generic", "representative" object. As such we attribute to a just those

---

[4] Put another way, although grammatically they are proper names in a language, they have no determinate reference (nor sense). For more on this, see (Milne, 2007, §§1 & 2).

properties possessed by all elements in the (notional) domain. With that in place, we have that $\phi(a) \in \Sigma_\infty$ if, and only if $\forall x \phi(x) \in \Sigma_\infty$. Consequently, $\Sigma_\infty \cup \{\forall x \phi(x)\} \vdash \chi$ if $\Sigma_\infty \cup \{\phi(a)\} \vdash \chi$. How does '$a$'s status as name for a *generic representative* manifest itself?—In our making no specific assumptions about a. So we obtain the rule

$$\begin{array}{c} [\forall x \phi(x)]^m \\ \vdots \\ \phi(a) \quad\quad \chi \\ \hline \chi \end{array} \; m \; \forall\text{-introduction}$$

where a does not occur in any premise upon which $\phi(a)$ depends.

In similar fashion we obtain the rules for the existential quantifier:

$$\begin{array}{c} [\exists x \phi(x)]^m \\ \vdots \\ \phi(t) \quad\quad \chi \\ \hline \chi \end{array} \; m \; \exists\text{-introduction.}$$

and

$$\begin{array}{c} [\phi(a)]^m \\ \vdots \\ \exists x \phi(x) \quad\quad \chi \\ \hline \chi \end{array} \; m \; \exists\text{-elimination}$$

where a occurs neither in χ nor in any side premise upon which χ depends.

These rules are all classically sound and, since we can obtain standard rules from them, they are complete. Moreover, the rules for $\wedge, \vee, \rightarrow, \neg$ and $\exists$ give us a classically complete system *with the subformula property* for the $\{\wedge, \vee, \rightarrow, \neg, \exists\}$-fragment of classical, first-order logic (Milne, 2010, § 3.3; Sandqvist, 2012).

As one quick example, the intuitionistically invalid $(\phi \rightarrow \psi) \rightarrow \psi \vdash \phi \vee \psi$ can be derived like this:

$$\dfrac{\dfrac{[\phi]^4 \quad [\phi \vee \psi]^2}{\phi \vee \psi} 2 \; \vee\text{-i} \quad\quad \dfrac{\dfrac{(\phi \rightarrow \psi) \rightarrow \psi \quad [\phi \rightarrow \psi]^4 \quad [\psi]^1}{\psi} 1 \; \rightarrow\text{-e} \quad [\phi \vee \psi]^3}{\phi \vee \psi} 3 \; \vee\text{-i}}{\phi \vee \psi} 4 \; \rightarrow\text{-i}.$$

Peter Milne

Addendum At Hejnice, Melvin Fitting asked whether we get anything new if we apply the line of argument developed here to intuitionist logic rather than classical. I said then that I suspected that one gets nothing essentially new. That is indeed the case. Perhaps the conditional will be sufficient illustration.

In the canonical Kripke model, nodes are prime theories and we have that

$$\phi \to \psi \notin \Sigma_\infty \text{ iff, for some prime theory } \Delta \text{ such that } \Sigma_\infty \subseteq \Delta, \\ \phi \in \Delta \text{ and } \psi \notin \Delta.$$

On the one hand, as in the classical case, we have, then, that if $\Sigma_\infty \cup \{\psi\} \vdash \chi$ then $\Sigma_\infty \cup \{\phi, \phi \to \psi\} \vdash \chi$ and we get again the general elimination form of *modus (ponendo) ponens*. On the other hand, given the Lindenbaum-Henkin construction in the completeness proof for intuitionist logic, a prime theory Δ such that $\Sigma_\infty \subseteq \Delta$, $\phi \in \Delta$ and $\psi \notin \Delta$ exists if, and only if, $\Sigma_\infty \cup \{\phi\} \nvdash \psi$. That $\Sigma_\infty \cup \{\phi \to \psi\} \vdash \chi$ and that $\Sigma_\infty \cup \{\phi\} \vdash \psi$ are, then, incoherent constraints. This leads to the rule

$$\begin{array}{cc} [\phi]^m & [\phi \to \psi]^m \\ \vdots & \vdots \\ \psi & \chi \\ \hline \multicolumn{2}{c}{\chi} \end{array} \; m \to.$$

which is nothing other than a rewriting of the standard $\to$-introduction rule.

3 An intriguing feature of classical logic

Our formulation of first-order classical logic does not have the subformula property. It does, however, satisfy this constraint: if $\Sigma \vdash \phi$ then there is a derivation of ϕ from Σ in which at most $\neg \phi$, subformulas of members of $\Sigma \cup \{\phi\}$, and negations of proper subformulas of members of $\Sigma \cup \{\phi\}$ occur (as follows from theorem 8 of (Milne, 2010, p. 210). This fact hints at—only hints at, every so light-handedly—a rather different treatment of negation, a treatment very much against the grain in proof-theoretic semantics but one which allows a uniform treatment of rules. (So much against the grain that in Carrollian spirit one might say its stands the proof-theoretic semantics/inferentialist account of negation on its head.)

We could get ourselves a whole lot more rules by noting facts such as this: $\neg(\phi \wedge \psi) \in \Sigma_\infty$ if, and only if, $\neg\phi \in \Sigma_\infty$ or $\neg\psi \in \Sigma_\infty$. Recast in terms of the Lindenbaum-Henkin condition of non-membership, we obtain:

- if $\Sigma_\infty \cup \{\neg\phi\} \vdash \chi$ and $\Sigma_\infty \cup \{\neg\psi\} \vdash \chi$ then $\Sigma_\infty \cup \{\neg(\phi \wedge \psi)\} \vdash \chi$;
- if $\Sigma_\infty \cup \{\neg(\phi \wedge \psi)\} \vdash \chi$ then $\Sigma_\infty \cup \{\neg\phi\} \vdash \chi$;
- if $\Sigma_\infty \cup \{\neg(\phi \wedge \psi)\} \vdash \chi$ then $\Sigma_\infty \cup \{\neg\psi\} \vdash \chi$.

These give us the rules:

$$\frac{\neg\phi \quad \begin{array}{c}[\neg(\phi\wedge\psi)]^m \\ \vdots \\ \chi\end{array}}{\chi} \; m \; \neg\wedge\text{-introduction (l)}; \qquad \frac{\neg\psi \quad \begin{array}{c}[\neg(\phi\wedge\psi)]^m \\ \vdots \\ \chi\end{array}}{\chi} \; m \; \neg\wedge\text{-introduction (r)};$$

$$\frac{\neg(\phi\wedge\psi) \quad \begin{array}{c}[\neg\phi]^m \\ \vdots \\ \chi\end{array} \quad \begin{array}{c}[\neg\psi]^m \\ \vdots \\ \chi\end{array}}{\chi} \; m \; \neg\wedge\text{-elimination.}$$

We can proceed similarly for disjunction, the conditional, and the quantifiers, obtaining, for example,

$$\frac{\neg\phi(t) \quad \begin{array}{c}[\neg\forall x\phi(x)]^m \\ \vdots \\ \psi\end{array}}{\psi} \; m \; \neg\forall\text{-introduction.}$$

and

$$\frac{\neg\forall x\phi(x) \quad \begin{array}{c}[\neg\phi(a)]^m \\ \vdots \\ \psi\end{array}}{\psi} \; m \; \neg\forall\text{-elim.}$$

where a occurs neither in ψ nor in any premise other than $\neg\phi(a)$ upon which ψ depends.

Noting too that since $\neg\neg\phi \notin \Sigma_\infty$ if, and only if, $\phi \notin \Sigma_\infty$, we have the rules

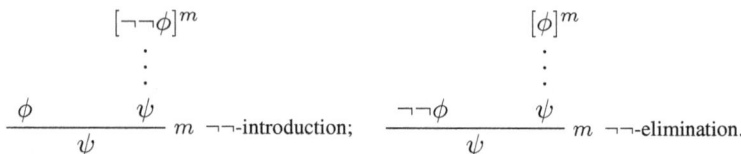

Albeit that they fall out of the Lindenbaum-Henkin conditions just as much as the rules above, these rules serve just as short cuts in a system with $\neg$-introduction and $\neg$-elimination. But these rules, together with the rules we already have for conjunction, disjunction, and the universal and existential quantifiers, *but not the rules for negation and the conditional*, all have a common structural feature: in the *introduction* rules, *all* side premises occur categorically; in the *elimination* rules, all side premises occur *hypothetically*.

And now, with negations in play, we can replace conditions such as $\phi \in \Sigma_\infty$ with $\neg\phi \notin \Sigma_\infty$, *i.e.*, $\Sigma_\infty \cup \{\neg\phi\} \vdash \chi$ to obtain rules of the same shapes:

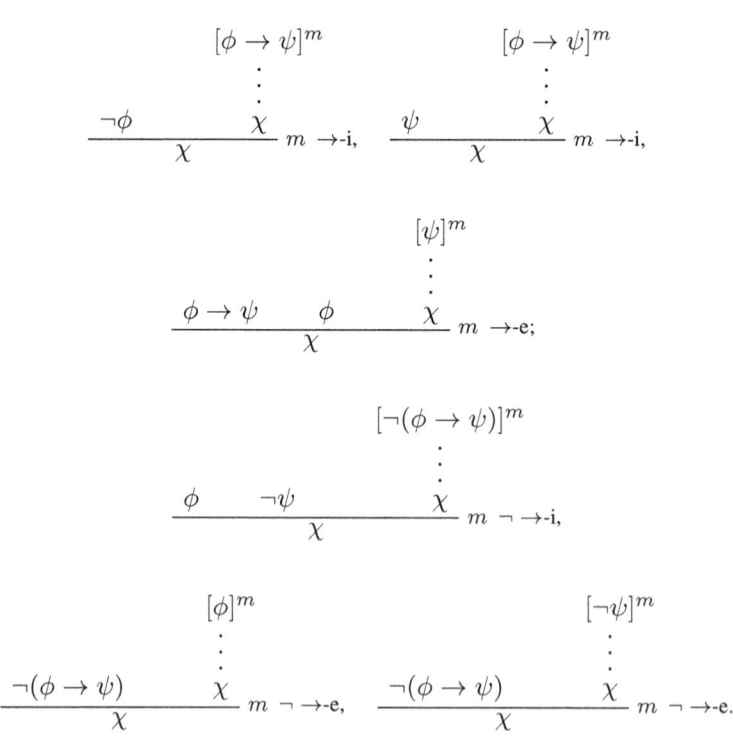

Classical Logic Through the Looking-glass

Now, rewrite *these* rules in the following ways.

Introduction rules Give each subformula ϕ_i a sequent to itself of the form $\Gamma_i \vdash \phi_i, \Delta_i$. These serve as the premises of right introduction rules. Give the formula introduced a sequent to itself of the form $\Gamma \vdash \star(\phi_1, \phi_2, \ldots), \Delta$ where Γ is the union of the antecedent side formulas Γ_i in premise sequents and Δ is the union of the succedent side formulas Δ_i in premise sequents. This serves as the conclusion of the rule.

Elimination rules Give each subformula ϕ_i a sequent to itself of the form $\Gamma, \phi_i \vdash \Delta_i$. These serve as the premises of left introduction rules. Give the formula eliminated a sequent to itself of the form $\Gamma, \star(\phi_1, \phi_2, \ldots) \vdash \Delta$ where, as before, Γ is the union of the antecedent side formulas in premise sequents and Δ is the union of the succedent side formulas in premises sequents. This serves as the conclusion of the rule.

For example, $\wedge$-introduction yields

$$\frac{\Gamma_1 \vdash \psi, \Delta_1 \qquad \Gamma_2 \vdash \phi, \Delta_2}{\Gamma_1, \Gamma_2 \vdash \psi \wedge \phi, \Delta_1, \Delta_2} \; R \wedge\text{-i},$$

$\wedge$-elimination gives us the pair of rules

$$\frac{\Gamma, \psi \vdash \Delta}{\Gamma, \psi \wedge \phi \vdash \Delta}, \quad \frac{\Gamma, \phi \vdash \Delta}{\Gamma, \psi \wedge \phi \vdash \Delta} \; L \wedge\text{-i}.$$

$\neg\wedge$-introduction gives us the pair of rules

$$\frac{\Gamma \vdash \neg\psi, \Delta}{\Gamma \vdash \neg(\psi \wedge \phi), \Delta}, \quad \frac{\Gamma \vdash \neg\phi, \Delta}{\Gamma \vdash \neg(\psi \wedge \phi), \Delta} \; R \neg\wedge\text{-i}.$$

$\neg\wedge$-elimination yields

$$\frac{\Gamma_1, \neg\psi \vdash \Delta_1 \qquad \Gamma_2, \neg\phi \vdash \Delta_2}{\Gamma_1, \Gamma_2, \neg(\psi \wedge \phi) \vdash \Delta_1, \Delta_2} \; L \neg\wedge\text{-i}.$$

The existential quantifier gives us

$$\frac{\Gamma, \psi(a) \vdash \Delta}{\Gamma, \exists x \psi \vdash \Delta} \; L \exists\text{-i}; \quad \frac{\Gamma \vdash \psi(t), \Delta}{\Gamma \vdash \exists x \psi, \Delta} \; R \exists\text{-i}$$

where the parametric name a does not occur in any member of Γ or Δ.

The negated existential quantifier gives us

$$\frac{\Gamma, \neg\psi(t) \vdash \Delta}{\Gamma, \neg\exists x \psi \vdash \Delta} \; L \neg\exists\text{-i}; \quad \frac{\Gamma \vdash \neg\psi(a), \Delta}{\Gamma \vdash \neg\exists x \psi, \Delta} \; R \neg\exists\text{-i}$$

where, again, the parametric name a does not occur in any member of Γ or Δ.

In these rules, by construction, the subformulas of interest and the formula containing the (left or right) introduced connective/quantifier all occur on the same side of the turnstile. Not so in the case of the standard sequent-calculus negation rules:

$$\frac{\Gamma \vdash \psi, \Delta}{\Gamma, \neg\psi \vdash \Delta} \text{ L } \neg\text{-i}; \quad \frac{\Gamma, \psi \vdash \Delta}{\Gamma \vdash \neg\psi, \Delta} \text{ R } \neg\text{-i}.$$

But as induction on length of proof shows, any sequent derived using the left negation rule can be obtained without it when we include axioms of the form $\phi, \neg\phi \vdash$, where ϕ is atomic; likewise, any sequent derived using the right negation rule can be obtained without it when we include axioms of the form $\vdash \phi, \neg\phi$, where ϕ is atomic.

With axioms of the forms $\phi \vdash \phi$ and $\neg\phi \vdash \neg\phi$, ϕ atomic, what we have here is the *cut-free* sequent calculus $\mathscr{S}'$ of (Smullyan, 1968). Smullyan (1968, Ch. XV, § 1) uses this calculus in proving first the Craig then the Craig-Lyndon interpolation theorem for first-order logic (without identity and function symbols). By attending very carefully to all the steps in Smullyan's proof, we obtain a refinement of the interpolation theorem for classical first-order logic. The first step towards that refinement is to note that when we drop axioms of the form $\vdash \phi, \neg\phi$ we have a cut-free sequent calculus for the first-order variant of Kleene's strong three-valued logic (*e.g.*, van Benthem, 1988, Avron, 1991, 2003, Busch, 1993); the second, to note that when, instead, we drop axioms of the form $\phi, \neg\phi \vdash$ we have a cut-free sequent calculus for Graham Priest's Logic of Paradox (Avron, 1991, 2003).

Exactly because of what we noted about where the formulas of interest stand with respect to the turnstile, we have immediately that $\Gamma \vdash$ iff $\Gamma \vdash_{K3}$ and $\vdash \Delta$ iff $\vdash_{LP} \Delta$. And if $\Gamma \vdash \Delta, \Gamma \nvdash$ and $\nvdash \Delta$ then at least one axiom, either of the form $\phi \vdash \phi$ or of the form $\neg\phi \vdash \neg\phi$, must be used. In a derivation of $\Gamma \vdash \Delta$, we associate ϕ with the axiom $\phi \vdash \phi$, $\neg\phi$ with the axiom $\neg\phi \vdash \neg\phi$; we associate nothing with axioms of the forms $\phi, \neg\phi \vdash$ and $\vdash \phi, \neg\phi$. We then push the interpolants downwards, changing them as appropriate given the rule employed. The aim is to associate interpolants with all and only those sequents $\Gamma' \vdash \Delta'$ in the derivation such that $\Gamma' \nvdash$ and $\nvdash \Delta'$.

For example, the interpolant, if there is one, is unchanged by L ∧-i, R ∨-i, L ¬∨-i, R ¬∧-e, L ¬¬-i, R ¬¬-i, L ∀-i, R ∃-i, L ¬∃-i, R ¬∀, and by Weakening/Augmentation, and none is introduced if there isn't one associated with the premise sequent.

The other rules give rise to less straightforward stipulations. I'm not

Classical Logic Through the Looking-glass

going to go through all of them. Here are two:

R ∧-i Let η be such that $\Gamma_1 \vdash_{K3} \eta$ and $\eta \vdash_{LP} \phi, \Delta_1$ and only non-logical vocabulary common to both Γ_1 and $\Delta_1 \cup \{\phi\}$ occurs in η; let θ be such that $\Gamma_2 \vdash_{K3} \theta$ and $\theta \vdash_{LP} \psi, \Delta_2$ and only non-logical vocabulary common to both Γ_2 and $\Delta_2 \cup \{\psi\}$ occurs in θ. By R ∧-i, $\Gamma_1, \Gamma_2 \vdash_{K3} \eta \wedge \theta$. By R ∧-i, $\eta, \theta \vdash_{LP} \phi \wedge \psi, \Delta_1, \Delta_2$; by two applications of L ∧-i, $\eta \wedge \theta \vdash_{LP} \phi \wedge \psi, \Delta_1, \Delta_2$. And only non-logical vocabulary common to both $\Gamma_1 \cup \Gamma_2$ and $\Delta_1 \cup \Delta_2 \cup \{\phi \wedge \psi\}$ occurs in $\eta \wedge \theta$.

If no interpolant is associated with $\Gamma_1 \vdash \phi, \Delta_1$ but $\Gamma_2 \vdash_{K3} \theta$ and $\theta \vdash_{LP} \psi, \Delta_2$ and only non-logical vocabulary common to both Γ_2 and $\Delta_2 \cup \{\psi\}$ occurs in θ, then, (i) if $\Gamma_1 \vdash$, no interpolant is associated with $\Gamma_1, \Gamma_2 \vdash \phi \wedge \psi, \Delta_1, \Delta_2$ and $\Gamma_1, \Gamma_2 \vdash$ by weakening and (ii) if $\vdash \phi, \Delta_1$ then θ is taken as the interpolant, for we have both that $\Gamma_1, \Gamma_2 \vdash_{K3} \theta$ and that $\theta \vdash_{LP} \phi \wedge \psi, \Delta_1, \Delta_2$. Likewise, *mutatis mutandis*, if no interpolant is associated with $\Gamma_2 \vdash \psi, \Delta_2$ but there is one associated with $\Gamma_1 \vdash \phi, \Delta_1$.

If no interpolant is associated with $\Gamma_1 \vdash \phi, \Delta_1$ and none with $\Gamma_2 \vdash \psi, \Delta_2$, none is associated with $\Gamma_1, \Gamma_2 \vdash \phi \wedge \psi, \Delta_1, \Delta_2$. If $\Gamma_i \vdash$ then $\Gamma_1, \Gamma_2 \vdash$ by Weakening, $i = 1, 2$. If $\vdash \phi, \Delta_1$ and $\vdash \psi, \Delta_2$ then $\vdash \phi \wedge \psi, \Delta_1, \Delta_2$ by R ∧-i.

L ¬∀-i Let $\Gamma, \neg\phi(a) \vdash_{K3} \eta$ and $\eta \vdash_{LP} \Delta$ where the parameter a does not occur in any member of Γ or Δ and only non-logical vocabulary common to both $\Gamma \cup \{\neg\phi(a)\}$ and Δ occurs in η. The parameter a does not occur in η hence, by L ¬∀-i, $\Gamma, \neg\forall x\phi \vdash_{K3} \eta$ and $\eta \vdash_{LP} \Delta$.

If no interpolant is associated with the sequent $\Gamma, \neg\phi(a) \vdash \Delta$, where the parameter a does not occur in any member of Γ or Δ, then none is associated with $\Gamma, \neg\forall x\phi \vdash \Delta$. If $\Gamma, \neg\phi(a) \vdash$ then, by L ¬∀-i, $\Gamma, \neg\forall x\phi \vdash$ (as the parameter a does not occur in any member of Γ).

The difference between what I do and what Smullyan did is that where I associate nothing with axioms of the form $\phi, \neg\phi \vdash$ and $\vdash \phi, \neg\phi$, he associates, respectively, the constants **f** and **t**. This smooths out his stipulations for the awkward rules. As indicated by the above, I have to be quite careful with what I stipulate for some of the connectives, quantifiers, and combinations of these with negation.

Matters are so set up that if an interpolant ϕ is associated with a sequent $\Gamma \vdash \Delta$ then ϕ can be derived from Γ *without appeal to an axiom of the form* $\vdash \psi, \neg\psi$ and Δ can be derived from ϕ *without appeal to an axiom of the form* $\psi, \neg\psi \vdash$. In other words, with '$\vdash$' as classical consequence, if $\Gamma \vdash \Delta$, then

(i) $\Gamma \vdash_{K3}$, or

(ii) $\vdash_{LP} \Delta$, or

(iii) $\Gamma \nvdash$, $\nvdash \Delta$, and there's an interpolant ϕ such that $\Gamma \vdash_{K3} \phi$ and $\phi \vdash_{LP} \Delta$,

where being an interpolant means that ϕ contains only non-logical vocabulary common to Γ and Δ. (Lyndon's parity constraints are also satisfied. In (Milne, in press-b) I take a related approach: I use block tableaux and extend the result to classical first-order logic with identity (but not function symbols). In (Milne, in press-a) I give a semantic proof of the analogous result for classical propositional logic.)

We have that in classical logic, without identity and function-symbols, derivations can be limited to three kinds: those in which the premises are classically and so $K3$-inconsistent, those in which the conclusion is classically and so LP-logically true, and the rest in which there is an interpolant such that the first part of the derivation comprises a $K3$ derivation from antecedent to interpolant and the second part comprises an LP derivation from interpolant to succedent.

The use of the negated rules is essential to the line of argument here but the result stands however one has formulated classical logic (without identity and function-symbols). However, neither $K3$ nor LP has a straightforward natural deduction formulation so this fact about classically valid sequents tends to be lost from view.

4 A contraction-free approximation to classical logic

I really did come at what I'm about to describe backwards—completely backwards. This section owes its origin to Francesco Tonci Ottieri, a student in my undergraduate logic class at the University of Stirling in the Spring Semester in 2015. He came to see me one day with a diagrammatic account of the semantics of indicative conditionals. It took us—him and me, independently—a while to realise that the diagrammatic aspect was but one way of assigning values to formulas subject to the sole constraint

Schema 1 $v(\phi \to \psi) = v(\phi) - v(\psi)$.

Under schema 1, $v(\phi \to \psi) + v(\psi \to \chi) = v(\phi \to \chi)$. We will follow Ottieri in taking this to vindicate *hypothetical syllogism*.[5]

We'll call this

Schema 2 $v(\phi \to \psi) = v(\psi) - v(\phi)$.

While schema 1 and schema 2 do just as good a job with hypothetical syllogism, schema 2, but not schema 1, gives us that $v(\phi) + v(\phi \to \psi) = v(\psi)$. This we take as vindicating *modus (ponendo) ponens*.

Should we want our conditionals to contrapose and, more exactly, more demandingly, should we want $\phi \to \psi$ and $\neg\psi \to \neg\phi$ to take the same value, both schemata give us

for any formulas ϕ and ψ, $v(\phi) + v(\neg\phi) = v(\psi) + v(\neg\psi)$.

The *only* way for this to work out is that, for some constant value K and all formulas ϕ,

$$v(\neg\phi) = K - v(\phi).$$

This gives us, as a consequence, that $v(\neg\neg\phi) = v(\phi)$.

Introducing a *(falsum)* constant, $\bot$, to which K is assigned as value, schema 2, but not schema 1, then gives us

$$v(\neg\phi) = v(\phi \to \bot),$$

leading to the familiar identification of $\neg\phi$ with $\phi \to \bot$.

Both schemata give us that $v(\phi \to \phi) = 0$, suggesting that 0 has some special role to play.

Schema 2, but not schema 1, gives us this:

$$v(\phi_1) + v(\phi_2) + \ldots + v(\phi_n) + v(\psi) \gtreqless v(\chi)$$

(if, and) only if

$$v(\phi_1) + v(\phi_2) + \ldots + v(\phi_n) \gtreqless v(\psi \to \chi).$$

Once we sort out a little detail, we'll read this as vindicating Conditional Proof/$\to$-introduction.

Given what schema 2 gets us, and what schema 1 doesn't, we'll adopt schema 2 as the preferred evaluation scheme for conditionals for the time being: for any formulas ϕ and ψ, $v(\phi \to \psi) = v(\psi) - v(\phi)$.

---

[5] Ottieri employs Schema 1 as the basis of his *algebraic demonstration* in (Ottieri, in press).

Peter Milne

Classically, $\phi \wedge \psi$ is equivalent to $\neg(\phi \to \neg\psi)$. $v(\neg(\phi \to \neg\psi)) = K - ((K - v(\psi)) - v(\phi)) = v(\phi) + v(\psi)$.

Setting $v(\phi \wedge \psi) = v(\phi) + v(\psi)$, we have, for all formulas ϕ, that $v(\phi \wedge \neg\phi) = K$. On the model of hypothetical syllogism and *modus ponens* above, setting $v(\phi \wedge \psi) = v(\phi) + v(\psi)$ vindicates $\wedge$-introduction. We also have that $v(\phi \wedge \psi) \geq v(\phi)$ and that $v(\phi \wedge \psi) \geq v(\psi)$.

These last two suggest that we should take as our criterion of validity: the inference from $\phi_1, \phi_2, \ldots, \phi_n$ to ψ is valid, which we write as $\phi_1, \phi_2, \ldots, \phi_n \Vdash \psi$, if, and only if, under every valuation v,

$$v(\phi_1) + v(\phi_2) + \ldots + v(\phi_n) \geq v(\psi).^{6,7}$$

From this we get Gentzen's rule for $\to$-introduction: if $\phi_1, \phi_2, \ldots, \phi_m \Vdash \psi$ and $\chi_1, \chi_2, \ldots, \chi_n, \tau \Vdash \upsilon$ then $\phi_1, \phi_2, \ldots, \phi_m, \chi_1, \chi_2, \ldots, \chi_n, \psi \to \tau \Vdash \upsilon$, for if $v(\phi_1) + v(\phi_2) + \ldots + v(\phi_m) \geq v(\psi)$ and $v(\chi_1) + v(\chi_2) + \ldots v(\chi_n) + v(\tau) \geq v(\upsilon)$ then $v(\phi_1) + v(\phi_2) \ldots v(\phi_m) + v(\chi_1) + v(\chi_2) + \ldots v(\chi_n) + (v(\tau) - v(\psi)) \geq v(\upsilon)$.

Commutativity of addition gives us the structural rule permutation (exchange) in the antecedent.

Transitivity/Cut falls out, for if

$$v(\phi_1) + v(\phi_2) + \ldots + v(\phi_n) + v(\chi) \geq v(\tau)$$

and

$$v(\psi_1) + v(\psi_2) + \ldots + v(\psi_n) \geq v(\chi)$$

then—obviously!—

$$v(\phi_1) + v(\phi_2) + \ldots + v(\phi_n) + v(\psi_1) + v(\psi_2) + \ldots + v(\psi_n) \geq v(\tau).$$

Weakening/Augmentation/Monotonicity falls out *if we insist that formulas take only non-negative values*. This gives 0 its special status, as the minimum value, and since, then, for any formula ϕ, $K - v(\phi) = v(\neg\phi) \geq 0$, K is the maximum attainable value. We obtain the standard $\bot$-elimination

---

[6]This might put the reader in mind, as it does me, of Dorothy Edgington's uncertainty semantics for classical logic augmented with Adams' conditionals (Edgington, 1992): there an inference is valid if under no probability distribution is the sum of the uncertainties of the premises less than the uncertainty of the conclusion (where *uncertainty* = 1 - *probability*).

[7]Had we put this criterion for validity in place sooner, we could have weakened contraposition so as to require only that, for all ϕ and ψ, $v(\phi \to \psi) \geq v(\neg\psi \to \neg\phi)$ and obtained the same stipulation for negation.

Classical Logic Through the Looking-glass

rule, $\bot \Vdash \phi$, and *ex falso quodlibet* or *Explosion*, $\phi, \neg\phi \Vdash \psi$, for all formulas ϕ. (We could stipulate that $K = 1$ —but why bother? As long as $K > 0$ we get what we want and avoid trivialisation.)

From Augmentation *via* Conditional Proof we get the positive paradox of material implication: $\psi \Vdash \phi \to \psi$. Explosion and Conditional Proof get us the negative paradox: $\neg\phi \Vdash \phi \to \psi$.

Restricting to non-negative values forces a small change in the way we treat conditionals and consequently conjunctions.

Schema 3 $v(\phi \to \psi) = \max\{v(\psi) - v(\phi), 0\}$.

As a quick run through verifies, this makes no difference to any of the substantive claims made concerning conditionals under schema 2 and validity and its preservation above. It does draw what we're doing closer to the evaluation clauses for connectives of Łukasiewicz's infinite-valued logic, especially as these are presented in, for example, (Restall, 1994). The validity criterion is, however, non-standard and does give us something different.

If we use the classical equivalent $\neg\phi \to \psi$ to define (intensional) disjunction, we find that $\vee$-elimination fails (although obviously disjunctive syllogism is sound). On the other hand, under schema 3, *but not schema 2*, Łukasiewicz's definition of disjunction as $(\phi \to \psi) \to \psi$ gives us exactly the familiar (extensional) disjunction:

$$\begin{aligned}v((\phi \to \psi) \to \psi) &= \max\{v(\psi) - v(\phi \to \psi), 0\} \\ &= v(\psi) - v(\phi \to \psi) \\ &= v(\psi) - \max\{v(\psi) - v(\phi), 0\} \\ &= v(\phi) \text{ if } v(\phi) \leq v(\psi) \\ &= v(\psi) \text{ otherwise.}\end{aligned}$$

That is $v((\phi \to \psi) \to \psi) = \min\{v(\phi), v(\psi)\}$. We adopt this as our evaluation clause for disjunction. It vindicates the standard rules for disjunction.

What we end up with is a contraction-free "approximation" to classical propositional logic which includes the standard, intuitionist introduction and elimination rules for $\wedge$, $\vee$, $\to$ and $\neg$ and Double Negation Introduction and Double Negation Elimination; these rules, however, are read with *multisets*, not sets, of assumptions and those rules which permit discharge of an assumption in their application are restricted to allow discharge of *only one occurrence* of the assumption (including one in each branch in the case of

∨-elimination). It's obvious that

$$\phi \to (\phi \to \psi) \nvDash \phi \to \psi.$$

(And since we have taken what are usually thought of as intensional conjunction and extensional disjunction, we should expect some classical distributivity principles to fail—*e.g.*, $\phi \vee (\psi \wedge \chi) \nvDash (\phi \vee \psi) \wedge (\phi \vee \chi)$.)

The semantics can be given what, at first sight at least, seems a reasonably natural interpretation. We regard the numbers as measuring the "potential for introducing error" into our reasoning. Error permeates from occurrences of assumptions down to conclusions. *Each* time you use an assumption, you open up a channel through which error may permeate; discharging the assumption closes the channel. That, I think, is the picture. Steps of reasoning are vindicated if, of necessity, the conclusion has no greater potential for error than the sum of the error potentials of the assumptions on which it depends.[8,9]

5   Drawing the strands together

Our "ideal rules" are designed with proof of the completeness of classical first-order logic in mind. As far as that job is concerned, they are ideal. The elimination rules closely match Gentzen's. The introduction rules are different but ask yourself this: Where did Gentzen get his introduction rules from? He doesn't tell us but I think it's fair to say that he didn't pluck them out of thin air. He got them, I think, from intuitionism. Given the time and his interests, intuitionism was the obvious source for rules which are somehow supposed to display meaning in conditions for proof. But once we have the idea of reasoning with assumptions, we can look again at the role of rules of inference. There are two things one needs to know about logically complex assumptions: how to move on, having made one (elimination rules), and how to make do without them (introduction rules).

---

[8]There's some loose analogy here with *aggregation of risk* considerations drawn up against multi-premise closure of knowledge under competently deduced consequence—see, *e.g.*, (Hawthorne, 2004, p. 47). As far as I am aware, epistemologists haven't brought such considerations to bear against Contraction; they worry about the risk accruing to belief in premises, not the *uses* of those premises in deductions.

[9]At Hejnice, Walter Carnielli, Libor Běhounek, and Chris Fermüller, I think, all pointed out to me that one can carry out this exercise in a weaker structure than a closed interval of the non-negative real numbers under the standard ordering with 0 as one end-point. But partly because it's how, set on course by Ottieri, I came across it and partly because it has, or at least seems to have, this natural reading, I prefer to stick with the way just outlined.

Classical Logic Through the Looking-glass

Our rules for the $\{\wedge, \vee, \rightarrow, \neg, \exists\}$-fragment do have the subformula property. This property doesn't carry over when we add the rules for the universal quantifier, the "constant domains inference" $\forall x(\phi(x) \vee \psi(a)) \vdash \forall x \phi(x) \vee \psi(a)$ being an obvious source of trouble. It might not be a great hardship to read "$\forall x$" as "$\neg \exists x \neg$" but it would be good to have a better understanding of what it is about the universal quantifier that leads the "ideal rules" approach into trouble—if trouble it is. Rather than see it as a problematic feature of classical logic, we can see it as a prompt to rethink how to incorporate negation. Our starting point, in the search for rules ideally suited to proving completeness, gives us a lead, as we saw.

Those rules helped us see that classically valid first-order inferences divide into those with inconsistent premises, those with logically true conclusions, and the rest. *Very roughly*, when deriving one of the rest, you can proceed initially as though you don't care that truth and falsity are exhaustive, reach a "middle point", and proceed thereafter as though you do now care about that but not that truth and falsity are mutually exclusive.

Given a set Γ of formulas, let $\neg\Gamma$ be the set $\{\neg\phi : \phi \in \Gamma\}$. We have that

$$\Gamma \vdash_{K3} \neg\Delta \text{ if, and only if, } \Delta \vdash_{LP} \neg\Gamma.$$

So there's a sense in which the advocate of Kleene's three-valued logic can recapture every classically valid inference; and likewise the advocate of Priest's Logic of Paradox. But neither of these logics is proof-theoretically *nice*, at least not in any of the ways inferentialists have come to value. Having negation play a special role, as it does in Smullyan's calculus, may be a first step to addressing those misgivings but may also, I suppose, just be a case of swapping one mystery for another.

Gentzen's introduction and elimination rules, restricted by a "discharge policy" that allows formula-instances to be discharged only one at a time, where Gentzen allows any number of instances of the formula in question on which the intermediate conclusion depends to be discharged in an application of the rule, give us a logic weaker than classical. Restricted this way, the standard proof of the Law of Excluded Middle,

$$\cfrac{\cfrac{\cfrac{\cfrac{[\phi]^1}{\phi \vee \neg\phi}\text{ V-i} \quad [\neg(\phi \vee \neg\phi)]^2}{\cfrac{\bot}{\neg\phi}\text{ 1 }\neg\text{-i}}\text{ }\bot\text{-i}}{\phi \vee \neg\phi}\text{ V-i} \quad [\neg(\phi \vee \neg\phi)]^2}{\cfrac{\cfrac{\bot}{\neg\neg(\phi \vee \neg\phi)}\text{ 2 }\neg\text{-i}}{\phi \vee \neg\phi}\text{ } DNE,}\text{ }\bot\text{-i}$$

is blocked when we try to discharge both occurrences of the assumption $\neg(\phi \vee \neg\phi)$ at the same time. We can turn this into a sound proof—sound with respect to the "error potential" semantics—of the conditional $\neg(\phi \vee \neg\phi) \to (\phi \vee \neg\phi)$ but that's not quite the same thing—really, *not at all* the same thing. The structural rule of Contraction, *i.e.*, the liberal discharge policy, is important. But now, here's a question: What makes a rule *structural*?

Section 1's "general introduction rules" $\neg$-introduction (Dilemma) and $\to$-introduction (a) (Tarski's Rule) are not sound in the error potential semantics. This can be seen very straightforwardly because neither $\phi \vee \neg\phi$ nor $\phi \vee (\phi \to \psi)$ must take the value 0. If, within that framework, we wish to amend the evaluation clauses so as to make them sound, we are *forced* to two-valuedness. At this point, something remarkable happens. The validity criterion becomes equivalent to this:

> under every valuation the conclusion takes the value 0 when all the premises do.

With that being the case, occurrences of a formula greater than one in number in the premises cease to make a difference, *i.e.*, Contraction *is* now sound. For example, whereas we had $\phi, \phi \Vdash \phi \wedge \phi$ but *not* $\phi \Vdash \phi \wedge \phi$, we now have the latter. We can *derive* it as follows:

$$\cfrac{\phi \quad \cfrac{\cfrac{\cfrac{\cfrac{[\phi]^1 \quad [\phi]^2}{\phi \wedge \phi}\wedge\text{-i}}{\phi \to (\phi \wedge \phi)}\to\text{-i (c)} \quad [\phi \to (\phi \wedge \phi)]^1}{\phi \to (\phi \wedge \phi)}\text{ 1 }\to\text{-i (a)} \quad [\phi \to (\phi \wedge \phi)]^2}{\phi \to (\phi \wedge \phi)}\text{ 2 }\to\text{-i (a)}}{\phi \wedge \phi}\to\text{-e}$$

At the cost of failure of the subformula property, we can work around retention of the illiberal discharge policy—a structural feature—by exploiting what was introduced as an operational rule.

Do general introduction rules blur the boundary between the operational and the structural? I suspect that they do. If so, is that a bad thing? It's tempting to think that the answer must be yes but I'm not sure why.

* * *

I have to say that I find these explorations on the other side of the mirror intriguing. I rather think they must have implications for inferentialism regarding the meaning of logical connectives and quantifiers—proof-theoretic semantics as the topic is often called. I have to say too, though, that I am at present far from clear as to what exactly those implications might be. Ending on that note, I'll take my cue from Lewis Carroll:

> The Red Queen shook her head. "You may call it 'nonsense' if you like," she said, "but *I've* heard nonsense, compared to which that would be as sensible as a dictionary!"

References

Avron, A. (1991). Natural 3-valued Logics—Characterization and Proof Theory. *Journal of Symbolic Logic, 56*, 276–294.

Avron, A. (2003). Classical Gentzen-type Methods in Propositional Many-valued Logics. In M. Fitting & E. Orlowska (Eds.), *Beyond Two: Theory and Applications of Multiple-valued Logic* (pp. 117–155). Heidelberg: Physica Verlag.

Busch, D. R. (1993). A Sequent Axiomatization of Three-valued Logic with Two Negations. In L. M. Pereira & A. Nerode (Eds.), *Logic Programming and Non-monotonic Reasoning, Proceedings of the Second International Workshop, Lisbon, Portugal, June 1993* (pp. 476–494). Cambridge, Massachusetts: MIT Press.

Carroll, L. (1970). *The Annotated Alice* (revised ed.; M. Gardner, Ed.). Harmondsworth: Penguin.

Edgington, D. (1992). Validity, Uncertainty and Vagueness. *Analysis, 52*, 193–204.

Fine, K. (1985). *Reasoning with Arbitrary Objects*. Hoboken: Blackwell Publishing.

Hawthorne, J. (2004). *Knowledge and Lotteries*. Oxford: Oxford University Press.

Milne, P. (2007). Existence, Freedom, Identity, and the Logic of Abstractionist Realism. *Mind*, *116*, 23–53.

Milne, P. (2008). A Formulation of First-order Classical Logic in Natural Deduction with the Subformula Property. In M. Peliš (Ed.), *Logica Yearbook 2007* (pp. 97–110). Prague: Filosofia.

Milne, P. (2010). Subformula and Separation Properties in Natural Deduction *via* Small Kripke Models. *Review of Symbolic Logic*, *3*, 175–227.

Milne, P. (2015). Inversion Principles and Introduction Rules. In H. Wansing (Ed.), *Dag Prawitz on Proofs and Meaning* (pp. 189–224). Berlin: Springer.

Milne, P. (in press-a). A Non-classical Refinement of the Interpolation Property for Classical Propositional Logic. *Logique & Analyse*.

Milne, P. (in press-b). A Refinement of the Craig-Lyndon Interpolation Theorem for Classical First-order Logic (with Identity). *Logique & Analyse*.

Næss, A. (1968). *Four Modern Philosophers*. Chicago: University of Chicago Press. (English translation by Alastair Hannay of Norwegian original.)

Ottieri, F. T. (in press). Logica, Ordine Geometrico Demonstrata. *British Journal of Undergraduate Philosophy*.

Restall, G. (1994). Arithmetic and Truth in Łukasiewicz's Infinitely Valued Logic (Tech. Rep. No. TR-ARP-6-94). Australian National University, Canberra: Automated Reasoning Project.

Sandqvist, T. (2012). The Subformula Property in Natural Deduction Established Constructively. *Review of Symbolic Logic*, *5*, 710–719.

Smullyan, R. M. (1968). *First-order Logic*. New York City: Springer. (Reprinted with corrections and a new preface, New York City: Dover Publications, 1995.)

van Benthem, J. (1988). *A Manual of Intensional Logic* (second ed.). Stanford: CSLI Publications.

Peter Milne
University of Stirling
Scotland
E-mail: peter.milne@stir.ac.uk

Incompatibility and Inference as Bases of Logic

JAROSLAV PEREGRIN[1]

Abstract: That logic can be based merely either on the concept of inference, or on that of incompatibility has been already shown. The question is whether such austere foundations predetermine the kind of logic we reach in such a way. In this paper we show that in the case of logic based on inference the natural outcome is intuitionist logic, while we can reach also classical logic (if we sacrifice naturalness). However, in case of logic based on incompatibility the outcome is not really optional: the resulting logic is classical and there is no obvious way how to reach intuitionist logic.

Keywords: inference, incompatibility, classical logic, intuitionist logic

It is not too controversial to say that we can base logic solely on the concept of inference. To indicate how this can be done, let us introduce an *inference structure* as an ordered pair, $\langle S, \vdash \rangle$, where where S is a set (of "sentences" or "formulas") and $\vdash \in \mathcal{P}(S) \times S$ is a relation between subsets of S and elements of S ("the relation of inferability"), such that[2]

($\vdash$1) $X, A \vdash A$,

($\vdash$2) if $X, A \vdash B$ and $Y \vdash A$, then $X, Y \vdash B$.

Then we can define incompatibility—let us call it *pseudoincompatibility* (for it is defined on the basis on inference and opinions on how this definition is successful may differ) and denote it be the sign $\triangle$—in the following way:

$\triangle X \equiv_{\text{Def.}} X \vdash A$ for every A.

Then we can define conjunctions, disjunctions etc. in the way pioneered by Koslow (1992): an element B of S is called *a negation* of an element A of S iff the following two conditions hold

---

[1] Work on this paper has been supported by Research Grant No. 13-21076S of the Czech Science Foundation.

[2] Of course we could also consider "substructural" versions of inference structures based on rejecting some of these conditions. However, we do not do this in this paper.

(1) $\triangle A, B$,

(2) if $\triangle A, D$, then $D \vdash B$.

The first condition states that the negation of A is pseudoincompatible with A, whereas the second one states that it is the *minimal* element of S with this property: B is inferable from any other element of S which is also incompatible with A.[3] It follows that any two negations of A are equivalent in the sense that they are interinferable. Note also that in general the negation of a given element of S need not exist.

The situation is, of course, different when we take S to be generated from a basic vocabulary by means of some grammatical rule and if we introduce a specific negation sign producing, for every element A of S, its negation $\neg A$, like in common languages of logic. In such cases (1) and (2) directly stipulate the behavior of this new element of S:

($\neg$1) $\triangle A, \neg A$,

($\neg$2) if $\triangle A, D$, then $D \vdash \neg A$.

All other logical operators can be introduced in a similar vein

($\wedge$1) $A \wedge B \vdash A$ and $A \wedge B \vdash B$,

($\wedge$2) if $D \vdash A$ and $D \vdash B$, then $D \vdash A \wedge B$,

($\vee$1) if $A \vdash D$ and $B \vdash D$, then $A \vee B \vdash D$,

($\vee$2) if (if $A \vdash D$ and $B \vdash D$, then $E \vdash D$), then $E \vdash A \vee B$,

($\rightarrow$1) $A, A \rightarrow B \vdash B$,

($\rightarrow$2) if $A, D \vdash B$, then $D \vdash A \rightarrow B$.

Thus, Koslowian definitions give us every logically complex sentence as a minimum of a certain propositional function.

It might be interesting to consider a slight modification of Koslow's definition of disjunction:

---

[3]This presupposes that we read "$D \vdash B$", in effect, as "$B \leq D$".

Incompatibility and Inference as Bases of Logic

(∨1′) $A \vdash A \vee B$ and $B \vdash A \vee B$,

(∨2′) if $A \vdash D$ and $B \vdash D$, then $A \vee B \vdash D$.

This disturbs the uniformity of Koslow's logic in that disjunction is no longer defined as the *minimum* of a propositional function (but rather a *maximum* of one), but it gives our definition of logical operators a more explicitly algebraic flavor: conjunction can be seen as *supremum* and disjunction as *infimum*. Anyway, the logic we reach in this way (both in the Koslowian, and in the modified one), not surprisingly, is intuitionistic.

Is there a way of reaching also classical logic in terms of inference? Yes, there is; it is enough to modify the definition of negation in the following way:

(¬1) $\triangle A, \neg A$,

(¬2′) if $\triangle \neg A, D$, then $D \vdash A$.

This is a definition not so neat as the previous one, but it does yield us classical logic. This is easily seen, for now we have

if $\triangle \neg A, \neg \neg A$, then $\neg \neg A \vdash A$

as an instance of (¬2′), and as the antecedent is an instance of (¬1), we have the consequent

$\neg \neg A \vdash A$

which is nothing else than the intuitionistically notoriously invalid law of double negation elimination.

So the *natural* outcome of basing logic on inference is intuitionistic logic; but if we are willing to sacrifice naturalness, we can reach classical logic as well.

Now turn your attention to logic based on incompatibility. The framework in which we can study this kind of logic is that of an *incompatibility structure*, which is an ordered pair $\langle S, \bot \rangle$, where S is a set and $\bot \in \mathcal{P}(S)$ is a set of subsets of S, such that

(⊥) if $\bot X$ and $X \subseteq Y$, then $\bot Y$.

Here we can define inference—call it *pseudoinference* and denote it by ▷—as follows:

$X \triangleright A \equiv_{\text{Def.}} \perp Y, A$ implies $\perp Y, X$ for every Y.

How it is possible to define logical operators in this setting was shown by Brandom (2008):

($\neg$) $\perp \neg A, X$ iff $X \triangleright A$,

($\wedge$) $\perp X, A \wedge B$ iff $\perp X, A, B$.

It is easy to see that this definition gives us classical logic: as certainly $A \triangleright A$, ($\neg$) gives us

$\perp \neg A, A,$

and as it also gives us

if $\perp \neg A, \neg \neg A$, then $\neg \neg A \triangleright A$,

we have

$\neg \neg A \triangleright A.$

Now the question is: is the situation similar to the previous one, in that though the most *natural* outcome of basing logic on incompatibility is classical logic, it would be possible to reach also intuitionistic one?

In light of the previous considerations concerning logic based on inference, it may seem that we might be able to reach intuitionistic logic by replacing ($\neg$) by

($\neg'$) $\perp A, X$ iff $X \triangleright \neg A$.

This is what I conjectured in (Peregrin, 2011). Since then, however, I delved deeper into the problem and I have found out that this impression is wrong (see Peregrin, 2015). Even if we replace ($\neg$) by ($\neg'$), we still have classical logic. For suppose that

$\perp A, X.$

Then, according to ($\neg'$),

$X \triangleright \neg A,$

Incompatibility and Inference as Bases of Logic

and hence, unpacking the definition of ▷,

if $\bot \neg\neg A, \neg A$, then $\bot \neg\neg A, X$.

And as it is the case that

$\bot \neg\neg A, \neg A$

(for according to $(\neg')$, this is equivalent to $\neg A \triangleright \neg A$),

$\bot \neg\neg A, X$.

Hence the assumption $\bot A, X$ yields us $\bot \neg\neg A, X$, which is nothing else than $\neg\neg A \triangleright A$.

Why is this? What plays the crucial role is the definition of (pseudo)inference in terms of incompatibility.

First issue to realize is that (as I was reminded by Peter Milne), the incompatibilities of intuitionistic logic are *the same* as those of classical logic. This follows from the Glivenko Theorem which states that A is a theorem of classical logic iff $\neg\neg A$ is a theorem of intuitionistic one. It follows that $\neg A$ is a theorem of classical logic iff it is a theorem of intuitionistic one: for $\neg A$ is a theorem of classical logic iff $\neg\neg\neg A$ is a theorem of intuitionist logic, which in turn holds iff $\neg A$ is a theorem of intuitionist logic (as $\neg\neg\neg A \leftrightarrow \neg A$ is a theorem of intuitionistic logic). Further it follows that if $X \vdash A \wedge \neg A$, hence if $\triangle X$, in classical logic, then the same holds in intuitionist logic. The reason is that if $X \vdash A \wedge \neg A$ in classical logic, then, as the logic is compact, $X^* \vdash A \wedge \neg A$ for some finite subset X^* of X, then $\wedge(X^*) \vdash A \wedge \neg A$, where $\wedge(X^*)$ is the conjunction of the elements of X^*, then $\vdash \neg \wedge (X^*)$, and hence $\vdash \neg \wedge (X^*)$ also in intuitionist logic and thus $X \vdash A \wedge \neg A$ is also in intuitionist logic.

Hence starting from incompatibility we should not expect that we will be able to differentiate classical and intuitionist logic. But this does not yet explain why we get classical, rather than intuinionistic logic. Assume we base logic on inference: then we can define pseudoincompatibility, $\triangle$, and then pseudinference, ▶, based on this pseudoincompatibility:

$X \blacktriangleright A \equiv_{\text{Def.}} \triangle Y, A$ implies $\triangle Y, X$ for every Y,

It is not necessary that $\vdash$ and ▶ coincide; and indeed in intuitionistic logic they do not. However, if you start from incompatibilities, then the *only* way to define inference (at least the only minimally reasonable way I can think of) is to make ▷ directly into ⊢, which is the way of classical logic. There is no way to deviate ⊢ from ▷, as intuitionistic logic requires.

Why is it the case that in classical logic, but not in intuitionistic logic, $X \blacktriangleright A$ brings about $X \vdash A$? Suppose $X \blacktriangleright A$. This is to say that for all Y, if $\triangle Y, A$, then $\triangle Y, X$. As $\triangle Y, A$ iff $Y \vdash \neg A$ (both in intuitionistic and classical logic), it follows that if $Y \vdash \neg A$, then $\triangle Y, X$. Since $\neg A \vdash \neg A$, it further follows that $\triangle \neg A, X$. And in classical logic, though not in intuitionistic one, it follows that $X \vdash A$.

To sum up, we can conclude that while it is arguably possible to base logic both on the sole concept of inference, and on the sole concept of incompatibility, the nature of the ensuing logics is different. In the first case, the "natural way" leads us directly to intuitionist logic, but there is an optional alternative, the "not-so-natural" way which results into classical logic. On the other hand, if we take incompatibility as our base concept, we are on the way to classical logic. The reason is that there does not seem to be a way which would build inference out of incompatibility that would not already instill it with the essence of classical logic.

References

Brandom, R. (2008). *Between Saying and Doing: towards Analytical Pragmatism*. New York: Oxford University Press. (Appendix to Chapter V, cowritten with Aker, A.)

Koslow, A. (1992). *A Structuralist Theory of Logic*. Cambridge: Cambridge University Press.

Peregrin, J. (2011). Logic as Based on Incompatibility. In M. Peliš & V. Punčochář (Eds.), *The Logica Yearbook 2010* (pp. 158–167). London: College Publications.

Peregrin, J. (2015). Logic Reduced to Bare (Proof-theoretical) Bones. *Journal of Logic, Language and Information*, 24, 193–209.

Jaroslav Peregrin
Institute of Philosophy, Czech Academy of Sciences
The Czech Republic
E-mail: `jarda@peregrin.cz`

Minimalism, Reference, and Paradoxes

LAVINIA PICOLLO[1]

Abstract: The aim of this paper is to provide a minimalist axiomatic theory of truth based on the notion of reference. To do this, we first give sound and arithmetically simple notions of reference, self-reference, and well-foundedness for the language of first-order arithmetic extended with a truth predicate; a task that has been so far elusive in the literature. Then, we use the new notions to restrict the T-schema to sentences that exhibit 'safe' reference patterns, confirming the widely accepted but never worked out idea that paradoxes can be characterised in terms of their underlying reference patterns. This results in a strong, ω-consistent, and well-motivated system of disquotational truth, as required by minimalism.

Keywords: minimalism, disquotation, reference, paradoxes, well-foundedness

1 Introduction

The core of minimalism, one of the most popular versions of deflationism about truth nowadays, consist of the following two theses: first, that the meaning of the truth predicate is exhausted by the T-schema, this is,

$$T\ulcorner\varphi\urcorner \leftrightarrow \varphi, \qquad \text{(T-schema)}$$

where T stands for the truth predicate, φ is a sentence and $\ulcorner\varphi\urcorner$ a quotational name for it.[2] Second, that the truth predicate is just a logico-linguistic device that exists in the language solely to allow us to express certain things—mainly generalisations—we simply cannot express otherwise. The latter prompts the construction of 'logics' or axiomatic theories of truth. The former thesis

---

[1] I'm obliged to Eduardo Barrio, Volker Halbach, Hannes Leitgeb, Thomas Schindler, the Buenos Aires Logic Group, and the MCMP logic group for their extremely useful comments, suggestions, and corrections on previous stages of this work.

[2] Actually, Horwich (1998), the main exponent of minimalism, takes propositions to be truth bearers rather than sentences. In his account $\ulcorner\varphi\urcorner$ should be understood as a canonical name of the proposition expressed by φ.

suggests the instances of the T-schema—i.e. the T-biconditionals—as axioms.

Unfortunately, as is well-known, if the language is capable of self-reference and the underlying logic is classical, the full T-schema leads to paradox. For we can formulate a liar sentence λ, that "says of itself" that it's *untrue*. Thus, we have that

$$\lambda \leftrightarrow \neg T\ulcorner\lambda\urcorner, \tag{1}$$

which obviously contradicts the T-biconditional for λ. As a consequence, minimalists choose to let some T-biconditionals go, as follows:

> [...] the principles governing our selection of excluded instances are, in order of priority: (a) that the minimal theory not engender 'liar-type' contradictions; (b) that the set of excluded instances be as small as possible; and—perhaps just as important as (b)—(c) that there be a constructive specification of the excluded instances that is as simple as possible. (Horwich, 1998, p. 42)

Theories consisting exclusively of instances of the T-schema are called *disquotational*. The search for a constructive and encompassing policy for selecting jointly-consistent instances of this principle is what we call the *minimalist project*.

The task is not as easy as it may seem. The most natural option, namely letting the instances that lead to contradiction go, is not available, as McGee (1992) has shown. There is not one but many different maximal consistent sets of T-biconditionals, all of which are highly complex—not even arithmetically definable. A stricter criterion than mere consistency is needed.

Horwich himself puts forward a plausible restriction:

> The intuitive idea is that an instance of the equivalence [T-] schema will be acceptable, even if it governs a proposition concerning truth (e.g. "What John said is true"), as long as that proposition (or its negation) is grounded—i.e. is entailed either by the non-truth-theoretic facts, or by those facts together with whichever truth-theoretic facts are 'immediately' entailed by them (via the already legitimised instances of the equivalence schema), or ... and so on. (Horwich, 2005, p. 81)

However, he doesn't specify in which way we should understand 'grounded' or 'entailed'. Moreover, the notions of *grounding* (Kripke, 1975) and *dependence on non-truth-theoretic facts* (Leitgeb, 2005) that are available in the literature, even though they can lead to a unique set of acceptable instances of the T-schema, are far from supporting a constructive specification.

Perhaps the criterion that fares best so far is that of T-positiveness: only sentences in which the truth predicate occurs positively (i.e. under the scope of an even number of negation symbols) are allowed in the T-schema (Halbach, 2009). This is a recursive restriction that results in an ω-consistent powerful system when formulated over Peano arithmetic, called PUTB.[3] However, T-positiveness is a highly artificial restriction. It leaves out many intuitively harmless instances of the T-schema, and is inconsistent with appealing truth principles, like consistency and the fact that Modus Ponens and Conditional Proof preserve truth.

According to the orthodox view on paradoxes driven by Poincaré, Russell and Tarski, among others, semantic paradoxes and other pathological expressions are characterised by a common reference pattern, namely, *self-reference*. That certainly seems to be the case for liar sentences. This view has never been thoroughly investigated, mainly because of the elusiveness of a sound notion of reference for formal languages. If true, self-reference could be employed as a plausible restriction on the T-schema. Moreover, since reference has a syntactic vein, the resulting criterion could be in principle simple enough to give axiomatic disquotational theories.

However, Yablo (1985, 1993) challenged the orthodox view with a *prima facie* non-self-referential semantic paradox. This antinomy gave rise to a lively debate on its referential status that put in evidence the lack of sound and precise notions of reference and self-reference in the literature to assess paradoxes in formal languages (cf. Cook, 2006; Leitgeb, 2002). Until we come up with such notions, neither the orthodox view nor the referential status of Yablo's paradox can be evaluated properly.

The first goal of this paper is to remedy this situation. After some technical preliminaries in section 2, section 3 provides precise and intuitively appealing definitions of reference, and thus self-reference and well-foundedness, for formal languages of truth. As it turns out, according to

---

[3] PUTB can relatively interpret the Ramified Theory of Truth up to the ordinal ϵ_0, $RT_{<\epsilon_0}$, an axiomatic version of Tarski's hierarchy of semantic theories, and the Kripke-Fererman theory KF, an axiomatisation of Kripke's fixed-point semantic theory with the strong Kleene valuation scheme. In fact, it can be show that all three systems have the same proof-theoretic power. For an introduction to the systems and proofs of the quoted results see (Halbach, 2011), instead.

our definitions, the orthodox view is wrong, for Yablo's paradox isn't self-referential. Nonetheless, we show it is still possible to characterise the semantic paradoxes in terms of their referential patterns: they are all non-well-founded, as Horwich notices. This will become evident in section 4. Since the new notions are of a proof-theoretic nature, we employ them in the construction of an axiomatic theory given by well-founded T-biconditionals. We show that this system is sound and at least as strong as the best regarded axiomatic theories in the literature. Thus, in section 5 we conclude it's a good candidate for minimalism, the second and main aim of this note.

2 Technical preliminaries

Let $\mathcal{L}$ be the language of first-order Peano arithmetic (PA), with $\neg, \rightarrow, \forall$ and $=$ as primitive logical symbols. Formulae containing $\wedge, \vee, \leftrightarrow$ and $\exists$ are understood as abbreviations. $\mathcal{L}$ contains one individual constant 0, the successor function symbol S, and finitely many other function symbols for primitive recursive (p.r.) functions, to be specified. $\mathcal{L}$ has no predicate symbols besides identity. Other relation symbols such as $<$ are mere abbreviations. For each $n \in \omega$, the complex term given by n occurrences of S followed by 0 is the numeral of n, which we note $\bar{n}$. $\mathbb{N}$ is the standard model of $\mathcal{L}$, with ω as its domain.

$\mathcal{L}_T$, our language of truth, expands $\mathcal{L}$ with a new predicate symbol T for truth. PAT is the result of formulating PA in $\mathcal{L}_T$, taking all the instances of induction given by formulae of this language as axioms. If $\Gamma \subseteq \omega$, let $\langle \mathbb{N}, \Gamma \rangle$ be the expansion of $\mathbb{N}$ to $\mathcal{L}_T$, assigning Γ to T as its extension.

The expressions of $\mathcal{L}_T$ can be codified with natural numbers *à la* Gödel, so that $\mathcal{L}$ and its extensions can be understood as talking about these expressions and sequences (instead of numbers). Given a particular coding and an expression σ of $\mathcal{L}_T$, $\#(\sigma)$ is the code of σ and $\ulcorner \sigma \urcorner$ is the numeral of this code. We assume a standard coding, which is effective and monotonic.[4] Usually, we identify expressions with their codes, for perspicuity.

As is well known, for any $n \in \omega$ the (semi-)recursive subsets of ω^n can be defined in $\mathcal{L}$ and (weakly) represented in PA.[5] Let $ClTerm(v)$ represent the recursive set of closed terms of $\mathcal{L}_T$. If TH $\subseteq \mathcal{L}_T$ is a recursively axiomatisable system, $Bew_{\text{TH}}(v)$ *weakly* represents the set of its theorems. If TH is

---

[4] I.e. if a string of symbols σ occurs in another string σ', then $\#(\sigma) < \#(\sigma')$.

[5] Actually, this is possible already in Robinson arithmetic, a subsystem of PA. We use the latter for uniformity.

PA, we omit the subscript. We assume that all predicates $Bew_{TH}(v)$ satisfy Löb's derivability conditions (cf. Löb, 1955).

For any expression σ, let $\vec{\sigma}$ abbreviate $\sigma_1, \ldots, \sigma_n$. The diagonalisation function, that takes a formula $\varphi(v, \vec{v})$ and returns $\forall v(v = \ulcorner\varphi\urcorner \to \varphi)$, is represented in PA by $Diag(u, v)$. The evaluation function, that takes a term t of $\mathcal{L}_T$ and returns the numeral of the number it denotes, is also recursive and representable in PA by $val(u, v)$.

We assume $\mathcal{L}$ contains the following function symbols for p.r. functions, and PA their corresponding definitions: $\dot{\neg}v$ for the function that maps φ into $\neg\varphi$, $u(v/w)$ for the substitution function, that takes a formula φ and two terms t and s and replaces s in φ with t, and $\dot{v}$ for the numeral function that assigns to each number n its numeral $\bar{n}$. $\mathcal{L}$ cannot contain a function symbol for the evaluation function for its own terms, on pain of triviality. However, we write $u^\circ = v$ for the evaluation function as short for $val(u, v)$.

Let $\forall v(\psi(\ulcorner\varphi(\dot{v})\urcorner))$ abbreviate $\forall v(\psi(\ulcorner\varphi\urcorner(\dot{v}/\ulcorner u\urcorner)))$, which allows us to quantify over the free occurrences of v in $\varphi[v/u]$ when φ is between corner quotes. Also, let $\forall t\varphi$ abbreviate $\forall v(ClTerm(v) \to \varphi)$. As before, instead of $\forall t(\psi(\ulcorner\varphi\urcorner(t/\ulcorner v\urcorner)))$ we write $\forall t(\psi(\ulcorner\varphi(t)\urcorner))$ to quantify over terms within Gödel quotes.

Later it will become useful to have in mind the proof of the following well-known result.

Theorem 1 (Weak diagonal lemma) *For any formula $\varphi(v, \vec{v}) \in \mathcal{L}_T$ there is a formula $\psi(\vec{v}) \in \mathcal{L}$ s.t.*

$$\text{PAT} \vdash \psi(\vec{v}) \leftrightarrow \varphi(\ulcorner\psi(\vec{v})\urcorner, \vec{v})$$

Proof. The result of applying the diagonalisation function to

$$\forall u(Diag(v, u) \to \varphi(u, \vec{v}))$$

is the formula

$$\forall v(v = \ulcorner\forall u(Diag(v, u) \to \varphi(u, \vec{v}))\urcorner \to \forall u(Diag(v, u) \to \varphi(u, \vec{v}))) \quad (2)$$

Let a be the numeral of the Gödel code of (2). (2) is equivalent in PAT to

$$\forall u(Diag(\ulcorner\forall u(Diag(v, u) \to \varphi(u, \vec{v}))\urcorner, u) \to \varphi(u, \vec{v}))$$

which is equivalent to $\varphi(a, \vec{v})$. $\square$

It's possible to strengthen this result using function symbols as follows:

Theorem 2 (Strong diagonal lemma)   *For any formula $\varphi(v, \vec{v})$ of $\mathcal{L}_T$ there is a term t s.t.*

$$\text{PA} \vdash t = \ulcorner \varphi(t, \vec{v}) \urcorner$$

It is commonly thought that both diagonal lemmata deliver self-referential expressions. For instance, applying strong diagonalisation to the predicate $\neg Bew(v)$ we obtain a term g s.t.

$$\text{PA} \vdash g = \ulcorner \neg Bew(g) \urcorner \qquad (3)$$

$\neg Bew(g)$ is a Gödel sentence of PA and it is usually understood as "saying of itself" that it isn't provable in PA. As is well known, this sentence is true and therefore unprovable in PA.

Finally, recall that formulae in $\mathcal{L}$ can be classified according to their quantificational—also called *arithmetical*—complexity into sets Σ_n, Π_n and $\Delta_n \subseteq \mathcal{L}$, with $n \in \omega$. These sets constitute the *arithmetical hierarchy*. If φ is logically equivalent to a formula where all quantifiers are bound, φ is both Σ_0 and Π_0. If φ is logically equivalent to a formula of the form $\forall \vec{v} \psi$, where $\psi \in \Sigma_n$, then $\varphi \in \Pi_{n+1}$. If φ is logically equivalent to a formula of the form $\neg \forall \vec{v} \psi$ where $\psi \in \Pi_n$, then $\varphi \in \Sigma_{n+1}$. Finally, if φ is both Π_n and Σ_n, we say that $\varphi \in \Delta_n$. Note that the sets in the hierarchy are cumulative, for it's always possible to add superfluous quantifiers at the beginning of a formula.

Recursive sets can be defined in $\mathcal{L}$ by Δ_0-formulae, and semi-recursive sets by Σ_1-formulae. Non-semi-recursive sets can only be defined by more complex formulae, if at all. Every Δ_0-formula is decidable in PA. If $\varphi \in \Sigma_1$ is true in the standard model, then PA $\vdash \varphi$, this is, PA is Σ_1-complete. For other, more complex expressions, we have no guarantees.

3   Alethic reference

In this section we focus on the reference of sentences of $\mathcal{L}_T$ to sentences of the same language. This isn't just any kind of reference but reference *through the truth predicate* or, as we call it, *alethic reference*. Intuitively, an expression alethically refers to all sentences that syntactically fall, as it were, under the scope of the truth predicate. This will become clear soon. The notion we provide, is, as we show, of a low arithmetical complexity, though this doesn't come without costs.

Minimalism, Reference, and Paradoxes

A sentence in a first-order language can refer to an object either by mentioning it or by quantifying over it. In the first case, the expression must contain a term t that denotes the object. Since we're only interested in alethic reference, we have the following definition.

Definition 1 *Let φ and ψ be sentences of $\mathcal{L}_T$. φ refers by mention to ψ, or m-refers, for short, iff φ contains a subsentence Tt and* PA $\vdash t = \ulcorner\psi\urcorner$.

Note that if t actually denotes the code of ψ then PA will be able to prove it, for identity statements don't contain quantifiers. Definition 1 covers many cases, like the liar sentence that obtains applying the strong diagonal lemma to $\neg Tv$, that is,

$$\text{PA} \vdash l = \ulcorner \neg Tl \urcorner, \tag{4}$$

that intuitively m-refers to itself. In general, any sentence that result from strongly diagonalising formulae that contain Tv as a subformula will m-refer to themselves. On the other hand, if we strongly diagonalise formulae that don't satisfy this condition, we might not get self-referential expressions. For instance, diagonalising $T\neg v$ we get

$$\text{PA} \vdash l' = \ulcorner T\neg l' \urcorner. \tag{5}$$

$T\neg l'$ is an alternative liar sentence that doesn't refer to itself according to definition 1 but only to its negation. The latter is actually the self-m-referential one. This follows from (5) and the fact that $\neg T\neg l'$ contains $T\neg l'$ as a subsentence.

Sentences of $\mathcal{L}_T$ can also refer to other sentences by quantifying over them. For instance,

$$\forall x(Bew(x) \to Tx) \tag{6}$$

intuitively refers to all theorems of arithmetic, while

$$\forall x Tx \tag{7}$$

seems to refer to everything. Conditionals allow us to restrict reference by quantification. Thus, if a universal quantifier or a string of universal quantifiers is followed by a conditional expression, we would like to say that it refers to whatever satisfies the antecedent, and otherwise it refers to everything.

However, things are not so simple. In the first place, talking about satisfaction introduces too much complexity into our notion, for to know whether an arbitrary code satisfies a certain formula we would have to look into the set of arithmetically true statements, which is not arithmetically definable. Thus, we turn to the notion of *provability* instead. After all, what matters to avoid paradoxes is that we cannot *derive* a contradiction or an unsound claim. Consequently, the resulting notion of reference via quantification—or *q-reference*, for short—will be tied to a particular system, the system whose provability predicate we employ in the definition. We work in PA, but any extension of Robinson arithmetic works as well.

Secondly, recall we're only interested in alethic reference here, so what matters is what actually falls under the scope of T. While in (6) all theorems of arithmetic fall under the scope of T, in $\forall x(Bew(x) \to T\neg x)$ only their negations do. Analogously, in (7) all sentences fall under T but in $\forall xT\neg x$ only negations do. And the same can be said of more complex expressions. For instance, in $\forall x(Bew(x) \to \forall y(y = \neg x \to \neg Ty))$, again, only negations of PA's theorems fall under the scope of the truth predicate. Thus, we define q-reference recursively. Roughly, a universal expression q-refers to whatever its instances m- or q-refer to, unless the universal quantifier is followed by a conditional, in which case we consider only the instances given by numerals that provably satisfy the antecedent.

Finally, note that if quantification is restricted by a conditional expression in which the truth predicate occurs both in the antecedent and the consequent—e.g. $\forall x(Tx \to Tx)$, our theory has no means to know which sentences fall in the scope of T; since the idea is to axiomatise truth in terms of reference, not vice versa. Sentences of this kind could exhibit dangerous reference patterns without us knowing. Therefore, we just treat them as non-conditional expressions.

Now we turn to the formal definition of alethic q-reference.

Definition 2 *Let φ, ψ be sentences of $\mathcal{L}_T$. φ q-refers to ψ in PA iff T occurs in φ and one of the conditions 1-3 holds:*

1. $\varphi := \forall \vec{v} \chi$ and

    (a) $\chi := Tt$ or $\chi := \neg \delta$ and, for some $\vec{k} \in \omega$, $\chi[\vec{k}/\vec{v}]$ q-refers to ψ or has a new occurrence of Ts as a subsentence s.t. PA $\vdash s = \ulcorner\psi\urcorner$; or

    (b) $\chi := \delta \to \gamma$ and

        i. both δ and γ contain T and for some $\vec{k} \in \omega$, $\chi[\vec{k}/\vec{v}]$ q-refers to ψ or contains a new occurrence of Tt as a subsentence s.t. PA $\vdash t = \ulcorner\psi\urcorner$, or

ii. only γ (δ) contains T and there exist $\vec{k} \in \omega$ and $1 \leq i \leq n$ s.t. PA $\vdash \delta[\vec{k}/\vec{v}]$ ($\neg\gamma[\vec{k}/\vec{v}]$) and ($\delta \to \gamma$)$[\vec{k}/\vec{v}]$ *q-refers to* ψ *or contains a new occurrence of* Tt *as a subsentence s.t.* PA $\vdash t = \ulcorner\psi\urcorner$.

2. $\varphi := \neg\chi$ and χ q-refers to ψ.

3. $\varphi := \chi \to \delta$ and either χ or δ q-refer to ψ.

By *a new occurrence of* Tt *in* $\chi[\vec{k}/\vec{v}]$ in the above definition we mean that Tt occurs in the result of replacing all occurrences of Tt in χ with $0 = 0$ (or any sentence not containing T) and then instantiating the variables $\vec{v}$ with $\vec{k}$. This is needed to avoid cases of m-reference passing as cases of q-reference—e.g. in $\forall x T \ulcorner \lambda \urcorner$.

According to definition 2, the liar sentence λ introduced in (1) q-refers to itself, as well as all sentences that are obtained by weakly diagonalising a predicate $\varphi(v)$ containing Tv as a subformula. Looking at the proof of theorem 1, we see that the real form of these sentences is

$$\forall u(u = \ulcorner \forall v(Diag(u,v) \to \varphi(v))\urcorner \to \forall v(Diag(u,v) \to \varphi(v))) \qquad (8)$$

Applying the clause (b)ii. of definition 2 twice, we get that (8) is q-self-referential. But just like in the case of m-reference, if Tv isn't a subformula of $\varphi(v)$, our definition cannot guarantee that the weak diagonalisation of this predicate will be a self-referential expression.

Note that the notion of q-reference could clash with some of our intuitions. If $g = \ulcorner \neg Bew(g)\urcorner$ as in (3), strongly diagonalising the predicate $\forall x(x = y \land \neg Bew(g) \to \neg Tx)$ delivers a term l^* s.t.

$$\text{PA} \vdash l^* = \ulcorner \forall x(x = l^* \land \neg Bew(g) \to \neg Tx)\urcorner \qquad (9)$$

Since $\neg Bew(g)$ is true in the standard model, intuitively we would say $\forall x(x = l^* \land \neg Bew(g) \to \neg Tx)$ q-refers to itself. However, we're thinking about reference *in* PA, so this won't be the case. For PA cannot prove its own Gödel sentence, on pain of triviality. This is a direct consequence of adopting provability instead of satisfaction for defining reference. As we will see later, this issue can be circumvented to some extent.

Putting the notions of m- and q-reference together isn't enough to define reference *simpliciter*. Consider the following identities:

$$l_1 = \ulcorner Tl_2 \urcorner \qquad (10)$$
$$l_2 = \ulcorner \neg Tl_1 \urcorner.$$

This statements can be proved in PA by slightly tweaking theorem 2. Together, they give rise to a paradox akin to the liar. Sentences Tl_2 and $\neg Tl_1$ m-refer only to each other but, intuitively, also refer to themselves, though *indirectly*. Alethic reference is a transitive relation.

Definition 3 *Let φ, ψ be sentences of $\mathcal{L}_T$. φ directly refers to ψ in PA iff it m- or q-refers to ψ in PA.*

Definition 4 *A sequence of sentences $\chi_0, \ldots, \chi_n \in \mathcal{L}_T$, $n \in \omega$, is a chain of reference in PA iff, for each $i < n$, χ_i directly refers to χ_{i+1} in PA.*

Definition 5 *Let φ, ψ be sentences of $\mathcal{L}_T$. φ refers to ψ in PA iff there's a chain of reference in PA starting with φ and ending with ψ.*

According to this definition, both Tl_2 and $\neg Tl_1$ refer to themselves, as we wanted.

It's worth noticing that the notion of reference we present is not extensional but *hyperintensional*: there are logically equivalent sentences that don't refer to the same things. For instance, $0 = 0$ and $T\ulcorner\lambda\urcorner \vee \neg T\ulcorner\lambda\urcorner$ are logically equivalent but, while the former doesn't refer to anything, the latter refers to λ. Unlike grounding or dependence, reference is based at least partly on syntactic features of sentences and, therefore, extensionality fails.

The notion of reference we introduced can be used to define relevant reference patterns, such as the following two.

Definition 6 *A sentence $\varphi \in \mathcal{L}_T$ is self-referential in PA iff it refers to itself in PA.*

According to this definition, sentences such as λ in (1), $\neg Tl$ in (4) and Tl_2 and $\neg Tl_1$ in (10) turn out to be self-referential.

Definition 7 *A sentence $\varphi \in \mathcal{L}_T$ is well-founded in PA iff there is no indefinitely extensible chain of reference in PA starting with φ.*

Every self-referential expression is obviously non-well-founded. But there are also non-well-founded sentences that don't refer to themselves. Yablo's paradox (Yablo, 1985, 1993) consist of an infinite sequence of sentences, each of which says of the ones coming after that they are untrue. In $\mathcal{L}_T$, Yablo's sentences can be formalised as $\forall x > \bar{n} \neg Tv(x)$, where $v(v) = \ulcorner \forall x > \dot{v} \neg Tv(x) \urcorner$. This identity statement is provable in PA by strong diagonalisation, guaranteeing the existence of the list in our formal setting.

Minimalism, Reference, and Paradoxes

According to definitions 6 and 7, no sentence in the sequence is self-referential, though they are all non-well-founded. It can be shown that an ω-inconsistency follows from the set of T-biconditionals for sentences in Yablo's list, so the paradox is actually an ω-paradox (cf. Ketland, 2005). If our definitions are correct, this shows that the orthodox view on semantic paradoxes is mistaken: there are non-self-referential (ω-)paradoxes. But this doesn't spell doom to our approach, for semantic paradoxes could share a reference pattern other than self-reference; for instance, non-well-foundedness. Later we will see this is actually the case.

It's easily seen that m-reference is recursive. Since the only proper non-recursive notion involved in the definition of q-reference is the semi-recursive notion of provability, and it occurs only positively, q-reference is also semi-recursive. By a similar reasoning, direct reference, reference and self-reference are semi-recursive as well. Well-foundedness, on the other hand, is more complex. Nonetheless, all of these notions can be defined in $\mathcal{L}$ and most of them at least weakly represented in PA. This sets reference further apart from the usual notions of grounding and dependence, and is enough to allow our notion to play a role in a disquotational axiomatisation of truth.

Being q-reference strictly semi-recursive, PA can prove all positive cases, but some negative ones won't be provable. For instance, PA has no means to know that

$$\forall x(x = \ulcorner 0 = 0 \urcorner \to Tx) \tag{11}$$

does not q-refer to itself. That would mean PA knows that $\neg Bew(\ulcorner \forall x(x = \ulcorner 0 = 0 \urcorner \to Tx)\urcorner = \ulcorner 0 = 0\urcorner)$, this is, its own consistency. Since we want to be able to determine which sentences exhibit safe referential patterns to take them as instances of the T-schema, and (11) clearly does, we must add axioms to inform our theory of *some* negative cases of q-reference—by Gödel's theorem, it's impossible to have them all. The simplest principle we can add is

$$\forall x(Bew(\neg x) \to \neg Bew(x)) \tag{QR}$$

Since QR is true-in-$\mathbb{N}$, PA + QR, or QR(PA) for short, is ω-consistent. Given that PA knows that $\ulcorner \forall x(x = \ulcorner 0 = 0 \urcorner \to Tx)\urcorner \neq \ulcorner 0 = 0\urcorner$ and, therefore, that $Bew(\ulcorner \forall x(x = \ulcorner 0 = 0 \urcorner \to Tx)\urcorner \neq \ulcorner 0 = 0\urcorner)$, we can conclude in QR(PA) that $\neg Bew(\ulcorner \forall x(x = \ulcorner 0 = 0 \urcorner \to Tx)\urcorner = \ulcorner 0 = 0\urcorner)$, which means that (11) doesn't q-refer to itself.

4 Well-founded truth

In the previous section we provided formal proof-theoretic notions of alethic reference, self-reference, and well-foundedness for sentences of $\mathcal{L}_T$ in PA. The next step is to use them in the formulation of axiomatic disquotational theories of truth.

In the spirit of Horwich's (2005, p. 81) idea cited in the introduction, the most natural choice is to relativise the T-schema to the predicate $Wf(v) \in \mathcal{L}$ that defines well-foundedness in PA according to definition 7. However, this wouldn't result in a consistent system. Coming back to our example in (9), recall that $\forall x(x = l^* \wedge \neg Bew(g) \to \neg Tx)$ $(= l^*)$ doesn't refer to anything in PA, for PA $\nvdash Bew(\ulcorner \neg Bew(g) \urcorner)$. Moreover, QR(PA) can prove this, by internalising a proof of Gödel's theorem. Thus, QR(PA) $\vdash Wf(l^*)$. But, as it turns out, the T-biconditional for $\forall x(x = l^* \wedge \neg Bew(g) \to \neg Tx)$ leads directly to paradox. The reason is that this sentence is well-founded in PA but *not in* QR(PA), where it's actually self-referential.

To avoid this problem we restrict our attention to those sentences whose referenced expressions do not increase when we adopt more powerful systems. We call them *r-stable*. To formally characterise them, we need the following auxiliary notion:

Definition 8 *A sentence $\varphi \in \mathcal{L}_T$ is dr-stable iff all its subformulae of the form $\psi \to \chi$ where a free variable occurs in the scope of T and exactly one of ψ, χ contains T are s.t. the one not containing T is Δ_0.*[6]

For instance, $T\ulcorner \forall x(Bew(x) \to Tx) \urcorner$ and (11) are dr-stable, while

$$\forall x(Bew(x) \to Tx)$$

isn't, for $Bew(v) \notin \Delta_0$. If a dr-stable sentence φ doesn't directly refer to another sentence ψ in PA, φ cannot directly refer to ψ in a stronger theory either, since PA already decides all instances of Δ_0-formulae.

Definition 9 *A sentence $\varphi \in \mathcal{L}_T$ is r-stable iff it is dr-stable and refers only to dr-stable sentences.*

Thus, $T\ulcorner \forall x(Bew(x) \to Tx) \urcorner$ isn't r-stable, but (11) is, because it only refers to $0 = 0$. R-*un*stable expressions bear a certain analogy with blind

---

[6] By just considering Δ_0-expressions and not also their PA-equivalents we're leaving behind many sentences which have a stable direct reference. However, this doesn't matter for our purposes, since in the axioms of our truth system the restriction on the T-schema will be closed under PAT-equivalence.

truth ascriptions: in both cases we don't know what we are asserting and, *a fortiori*, if it's a paradox or not. Only for r-stable sentences we can be sure that their reference patterns are safe.

Since the set of Δ_0-expressions is obviously semi-recursive, so is the set of dr-stable sentences. Given that reference is also semi-recursive, r-stability has Π_2-complexity. Let $RSt(v) \in \Pi_2$ define this set. The theory we introduce next restricts the T-schema to r-stable and well-founded sentences and their equivalents *in a uniform way*.

Definition 10 WFUTB $\subseteq \mathcal{L}_T$ *extends* QR(PA) *with the new instances of induction for $\mathcal{L}_T$-formulae and the following schema, where $\varphi \in \mathcal{L}_T$ contains exactly n free variables:*

$$\forall \vec{t} \forall x (RSt(x(\vec{t})) \wedge Wf(x(\vec{t})) \wedge$$
$$\wedge Bew_{\text{PAT}}(\ulcorner\varphi(\vec{t})\urcorner \leftrightarrow x(\vec{t})) \rightarrow (T\ulcorner\varphi(\vec{t})\urcorner \leftrightarrow \varphi(\vec{t}^\circ)))$$

WFUTB—for *Well-founded Uniform Tarski Biconditionals*—allows instances of the T-schema given, uniformly, by all sentences that are equivalent in PAT to an r-stable well-founded sentence. This includes of course, all r-stable well-founded expressions, but also, for example, $\forall x((Tl \rightarrow Tl) \wedge x = \ulcorner 0 = 0 \urcorner \rightarrow Tx)$ and $\neg \forall x(Tx \rightarrow Tx)$, which are not well-founded in PA. On the other hand, it excludes many intuitively safe instances, such as the one given by $\forall x(Bew(x) \rightarrow Tx)$. We get the following results:

Proposition 1 WFUTB *is ω-consistent.*

Proof. We just give a sketch. It can be shown that if a dr-stable sentence $\varphi \in \mathcal{L}_T$ doesn't refer directly to another sentence ψ, then there's a set $\Gamma \subseteq \mathcal{L}_T$ on which φ depends s.t. $\psi \notin \Gamma$, by induction on the logical complexity of φ.[7] It follows as a corollary that all r-stable well-founded sentences belong to Leitgeb's set Φ_{lf} of expressions that depend on non-semantic states of affairs (cf. Leitgeb, 2005, § 3), by transfinite induction on the ordinal level of the fixed-point construction that leads to Φ_{lf}. Since there's a model $\langle \mathbb{N}, \Gamma \rangle$ of $\mathcal{L}_T$ that verifies all instances of the T-schema given by sentences in Φ_{lf} (Leitgeb, 2005, theorem 17), $\langle \mathbb{N}, \Gamma \rangle \vDash$ WFUTB as well. □

Proposition 2 *The theory of Ramified Truth up to ϵ_0 RT$_{<\epsilon_0}$ is relatively interpretable in* WFUTB.

---
[7]For a definition of *dependence* and its basic properties, see (Leitgeb, 2005).

Proof. We just give an idea of the proof.[8] We show that for each $\alpha < \epsilon_0$ there's a predicate $\theta_{\bar{\alpha}}(v) \in \mathcal{L}_T$ that satisfies in WFUTB the axioms that hold for $T_\alpha(v)$ in RT$_{<\epsilon_0}$.[9] First, we obtain a binary predicate $\theta_y(x) \in \mathcal{L}_T$ by strongly diagonalising over the variable w a complex predicate that is basically the disjunction of the axioms of RT$_{<\epsilon_0}$, where the predicates $T_\alpha(v)$ have been replaced by $Tw(\dot{y}/\ulcorner y \urcorner)(\dot{u}/\ulcorner x \urcorner)$ (and, correspondingly, α with y and v with u). Then we show by internal transfinite induction on α that the uniform T-schema holds in WFUTB for all predicates $\theta_{\bar{\alpha}}(v)$, where $\alpha < \epsilon_0$, which gives us the axioms of RT$_{<\epsilon_0}$. This is done by uniformly showing in WFUTB that all instances of the predicates $\theta_{\bar{\alpha}}(v)$ given by sentences in which only predicates $\theta_{\bar{\beta}}(v)$ with $\beta < \alpha$ occur are r-stable and well-founded. □

As a corollary of propositions 1 and 2, WFUTB is a sound and powerful system. Since the Kripke-Feferman theory KF and PUTB have the same proof-theoretic strength as RT$_{<\epsilon_0}$, WFUTB is at least as strong as these three well-regarded systems.

5 Conclusions

In this paper we have provided sound, precise, and arithmetically simple notions of reference, self-reference, and well-foundedness. Moreover, these concepts have been proved useful in the assessment of semantic paradoxes and in the formulation of axiomatic theories of truth.

We have also shown that a natural theory of disquotational truth that is ω-consistent, as powerful as KF and PUTB, and imposes only arithmetical restrictions on the T-schema is possible. Our system WFUTB is therefore (a) sound, (b) encompassing, and (c) employs a simple selective criterion of T-biconditionals. As a consequence, it's a perfect candidate for the minimalist search.

Perhaps other—more powerful—systems can be devised using the notions we introduced in section 3. It could well be that paradoxes shared more specific reference patterns than non-well-foundedness, which could be turned into broader selective criteria for instances of disquotation. We

---

[8] The proof is similar to the demonstration of Halbach's (2011, theorem 15.25).

[9] As is well known, natural numbers can codify ordinals up to ϵ_0 (and beyond). If $\alpha < \epsilon_0$, $\bar{\alpha}$ is the numeral of its code. PA is able to prove all instances of transfinite induction up to ϵ_0. For the details see (Pohlers, 2009, chapter 3).

believe this note not only provides answers to several issues such as finding a natural minimalist theory or assessing the orthodox view on semantic paradoxes, but also opens a new line of research on these topics.

References

Beall, J. C. (2005). Transparent Disquotationalism. In B. Armour-Garb & J. C. Beall (Eds.), *Deflationism and Paradox* (pp. 7–22). Oxford: Oxford University Press.
Cook, R. T. (2006). There Are Non-circular Paradoxes (but Yablo's Isn't One of Them!). *The Monist, 89*, 118–149.
Halbach, V. (2009). Reducing Compositional to Disquotational Truth. *Review of Symbolic Logic, 2*, 786–798.
Halbach, V. (2011). *Axiomatic Theories of Truth*. Cambridge: Cambridge University Press.
Horwich, P. (1998). *Truth* (second ed.). Hoboken: Blackwell Publishing.
Horwich, P. (2005). A Minimalist Critique of Tarski on Truth. In B. Armour-Garb & J. C. Beall (Eds.), *Deflationism and Paradox* (pp. 75–84). Oxford: Oxford University Press.
Ketland, J. (2005). Yablo's Paradox and ω-inconsistency. *Synthese, 145*, 295–307.
Kripke, S. (1975). Outline of a Theory of Truth. *Journal of Philosphy, 72*, 690–716.
Leitgeb, H. (2002). What Is a Self-referential Sentence? Critical Remarks on the Alleged (Non)-circularity of Yablo's Paradox. *Logique et Analyse, 45*(177–178), 3–14.
Leitgeb, H. (2005). What Truth Depends On. *Journal of Philosphical Logic, 34*, 155–192.
Löb, M. H. (1955). Solution of a Problem of Leon Henkin. *Journal of Symbolic Logic, 20*, 115–118.
McGee, V. (1992). Maximal Consistent Sets of Instances of Tarski's Schema. *Journal of Philosphical Logic, 21*, 235–241.
Pohlers, W. (2009). *Proof Theory: the First Step into Impredicativity*. Berlin: Springer.
Yablo, S. (1985). Truth and Reflexion. *Journal of Philosphical Logic, 14*, 297–349.
Yablo, S. (1993). Paradox without Self-reference. *Analysis, 53*, 251–252.

Lavinia Picollo

Lavinia Picollo
Ludwig-Maximilians University Munich
Germany
E-mail: `Lavinia.Picollo@lrz.uni-muenchen.de`

A Nonstandard Semantic Framework for Intuitionistic Logic

VÍT PUNČOCHÁŘ[1]

Abstract: In this paper, a nonstandard semantic framework for intuitionistic logic is introduced and its relation to Kripke semantics is studied. The main peculiarity of the framework is that it allows for information states that support a disjunction without supporting any of its disjuncts. The semantic structures of the framework are called information models and they consist of a join-semilattice with a zero element and a valuation assigning to every atomic formula an ideal in the algebraic structure. A method will be described that transforms any Kripke model into an equivalent information model and any information model into an equivalent Kripke model. It will also be shown how to extend the framework to the case of first-order logic.

Keywords: Intuitionistic logic, Kripke semantics, algebraic semantics, information states, distributive semilattices

1 Introduction

In this paper, we will describe a nonstandard semantic framework for intuitionistic logic which we will denote as SSIL (semilattice semantics for intuitionistic logic). The framework has some common features with Kripke semantics but it also differs from Kripke semantics in several important respects. Most importantly the framework allows for information states that support a disjunction without supporting any of its disjuncts.

A related framework for modal logics was introduced in (Punčochář, 2014). A similar framework viewed from the topological perspective was studied in (Punčochář, 2015) where it was presented as a generalization of inquisitive semantics (see Ciardelli & Roelofsen, 2011). Some mathematical aspects of the present framework, especially its relation to algebraic semantics of intuitionistic logic, are explored in detail in (Punčochář, submitted). The main goal of the present paper is to provide a careful comparison of SSIL and Kripke semantics for intuitionistic logic. We will show how

---

[1] Work on this paper was supported by grant no. P401/11/0371 of the Czech Science Foundation.

to transform any Kripke model into an equivalent model of SSIL, and how to transform any model of SSIL into an equivalent Kripke model. Mainly, we will work with the propositional language, but we will also sketch how to extend our approach to the case of first-order logic.

2 Algebras of information states

In this section, we will describe propositional SSIL and show some of its basic properties. We will work with a standard propositional language, which is built out of a set of atomic formulas At and a constant for contradiction $\bot \notin At$ using the connectives $\wedge, \vee, \rightarrow$. This language (or, more precisely, the set of its formulas) will be denoted as L. The Latin letters p, q will serve as variables for atomic formulas and the Greek letters φ, ψ as variables for arbitrary formulas. Negation might be defined as implication of contradiction: $\neg\varphi =_{def} \varphi \rightarrow \bot$. The semantic structures of SSIL are called algebras of information states.

Definition 1 *An algebra of information states is a join-semilattice with a zero element, i.e., an algebraic structure* $\mathcal{A} = \langle S, +, 0 \rangle$, *where S is a nonempty set (of information states), $+$ is an associative, commutative, and idempotent binary operation on S, and 0 is a neutral element with respect to $+$.*

Let $\mathcal{A} = \langle S, +, 0 \rangle$ be an algebra of information states. One can define an ordering on S in the following way:

$a \leq_\mathcal{A} b$ iff $a + b = b$.

The subscript will be omitted when no confusion arises. Associativity, commutativity, and idempotence of $+$ guarantee that $\leq$ is indeed a partial order. Moreover, $a + b$ can be recovered from $\leq$ as the least upper bound of $\{a, b\}$. To say that 0 is a neutral element corresponds to saying that 0 is the least element of the structure. So, algebras of information states can be alternatively seen as partially ordered sets with a least element and such that there is a least upper bound for any two elements.

The least element 0 represents an inconsistent state. The state $a + b$ represents a state consisting of information that is common to the states a and b. In this respect, our semantic framework differs from several similar semantics such as those introduced in (Veltman, 1984), (Urquhart, 1972), and (Fine, 2014). These semantics are also based on semilattices. However,

A Nonstandard Semantic Framework

the algebraic operation of these structures is interpreted in a different way. It is understood as a fusion of two states. A piece of information is supported by the fusion of a and b iff it is supported by a or by b. In contrast, in SSIL a piece of information is supported by the state $a+b$ iff it is supported by both a and b. A similar operation on information states is present in Wansing's informational semantics for substructural subsystems of Nelson logics (see Wansing, 1993).

The proposed interpretation of the two algebraic components, 0 and $+$, leads to the definition of a valuation. Every atomic formula is supported by the inconsistent state 0. Moreover, a given atomic formula p is supported by $a + b$ iff it is supported by a and b. In other words, the set of states that support p has to be an ideal in the algebra of information states.

Definition 2 *Let $\mathcal{A} = \langle S, +, 0 \rangle$ be an algebra of information states. A subset $I \subseteq S$ is an ideal in $\mathcal{A}$ if $0 \in I$ and the following holds:*

$a + b \in I$ iff $a \in I$ and $b \in I$.

With respect to the ordering, ideals are nonempty downward closed sets that are also closed under the operation $+$.

Definition 3 *An information model is a pair $\mathcal{M} = \langle \mathcal{A}, V \rangle$, where $\mathcal{A}$ is an algebra of information states and V a valuation defined as a function assigning an ideal to every atomic formula. We say that $\mathcal{M}$ is an information model on $\mathcal{A}$.*

Now, relative to an information model $\mathcal{M}$, we can define a relation of assertibility between information states and formulas of the language L. The expression '$a \Vdash \varphi$ in $\mathcal{M}$' is read as 'φ is assertible in a in the model $\mathcal{M}$'. If clear from the context, the reference to the model $\mathcal{M}$ will be omitted. The assertibility relation is defined by the following recursive conditions:

$a \Vdash \bot$ iff $a = 0$,

$a \Vdash p$ iff $a \in V(p)$,

$a \Vdash \varphi \wedge \psi$ iff $a \Vdash \varphi$ and $a \Vdash \psi$,

$a \Vdash \varphi \vee \psi$ iff there are b, c such that $b \Vdash \varphi$, $c \Vdash \psi$ and $b + c = a$,

$a \Vdash \varphi \to \psi$ iff for any $b \leq a$, if $b \Vdash \varphi$, then $b \Vdash \psi$.

One interesting observation is that when we make the semantics more concrete, and specify that information states are (arbitrary) sets of possible worlds, $+$ is union, and, as a consequence, $\leq$ is the inclusion relation, then we receive an informational semantics for classical logic. This semantics is used in the context of dependence logic (Yang, 2014) and inquisitive logic (Ciardelli, 2016). Our framework was developed as a natural generalization of this particular case.

It turns out that one more restriction imposed on algebras of information states is needed.

Definition 4 *An algebra of information states $\mathcal{A} = \langle S, +, 0 \rangle$ is distributive if it satisfies the following condition:*

(D) *if $a \leq b + c$, then there are d, e such that $d \leq b$, $e \leq c$ and $d + e = a$.*

It is well known that this condition generalizes standard distributivity of lattices: a lattice is distributive in the standard sense iff the corresponding join-semilattice is distributive in the sense of definition 4 (see Grätzer, 2011, p. 167).

The proposition expressed by φ in an information model $\mathcal{M}$ is the set of states of $\mathcal{M}$ in which φ is assertible. We will denote the proposition as $||\varphi||_\mathcal{M}$. Again, the subscript will usually be omitted.

It is shown in (Punčochář, submitted) that an algebra of information states $\mathcal{A}$ is distributive if and only if it holds for every information model $\mathcal{M}$ on $\mathcal{A}$ that every formula of L expresses an ideal. So, given a non-distributive algebra, one can always find a valuation so that there are formulas that do not express ideals. This is illustrated with the following typical non-distributive algebra of information states:

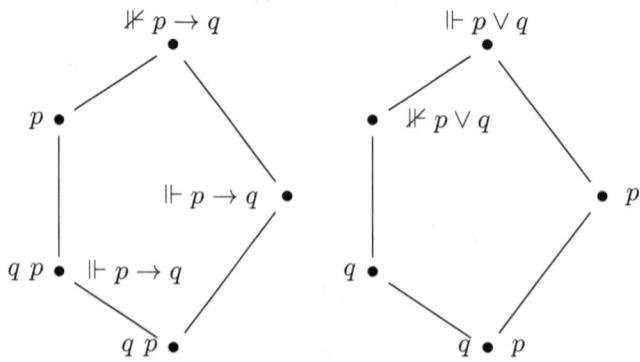

A Nonstandard Semantic Framework

The logic of all distributive information models is the intuitionistic logic. This is proved in (Punčochář, submitted). So, the proposed semantics SSIL can be regarded as a viable alternative to such frameworks as Kripke relational semantics and the standard algebraic semantics for intuitionistic logic. The relation of SSIL to algebraic semantics is surprisingly strong. It is studied in detail in (Punčochář, submitted), where it is shown that if $\mathcal{M}$ is distributive then the algebra of propositions is the Heyting algebra of ideals, i.e., the following holds for the algebra of ideals:

$||\bot||$ is the least ideal $\{0\}$.

$||\varphi \wedge \psi||$ is the greatest lower bound of $\{||\varphi||, ||\psi||\}$.

$||\varphi \vee \psi||$ is the least upper bound of $\{||\varphi||, ||\psi||\}$.

$||\varphi \to \psi||$ is the pseudocomplement of $||\varphi||$ relative to $||\psi||$.

Moreover, any Heyting algebra enriched with a valuation (in the sense of algebraic semantics) can be easily transformed into a distributive information model that validates the same formulas. Three different constructions of this kind are described in (Punčochář, submitted).

In (Punčochář, 2015), a less general version of the present semantics was introduced. In that paper, information states are defined as open sets of a topological space and formulas express principal ideals in the algebra of open sets. Several connections to Kripke semantics were described. We focused especially on the so called Alexandroff topologies since information models based on these topologies are closely related to Kripke models for intuitionistic logic. In the next section, we will study the connection to Kripke semantics from the perspective of the more general framework introduced in this paper.

3 A comparison with Kripke semantics

SSIL strongly reminds the Kripke semantics for intuitionistic logic (Kripke, 1965). The aim of this section is to point out some differences between the two frameworks as well as to reveal some important connections. Although SSIL can be regarded as a relational semantics, it will turn out that its relation to Kripke semantics is almost the same as the relation of algebraic semantics to Kripke semantics. We will start with an exposition of the basic concepts of Kripke semantics. The semantic structures are partially ordered sets. Valuation is usually defined as a function assigning an upward closed set to every atomic formula. For the sake of comparison, we will reverse the order, which means that valuations will assign downward closed sets. It is obvious that this modification does not make any significant difference.

Definition 5 *A Kripke frame is a pair* $\mathcal{F} = \langle S, \leq \rangle$, *where S is a nonempty set and $\leq$ is a partial order on S, i.e., a reflexive, transitive and antisymmetric relation. A Kripke model is a pair* $\mathcal{K} = \langle \mathcal{F}, V \rangle$, *where $\mathcal{F}$ is a Kripke frame and V a valuation defined as a function assigning to every atomic formula a downward closed set, i.e., if $a \in V(p)$ and $b \leq a$ then $b \in V(p)$.*

In the Kripke semantics for intuitionistic logic, these structures are interpreted (as in SSIL) as structures consisting of information states. Given a Kripke model $\mathcal{K}$, a relation (that might also be called a relation of assertibility) between formulas and information states is defined by the following recursive conditions:

$a \nvDash \bot$,

$a \vDash p$ iff $a \in V(p)$,

$a \vDash \varphi \wedge \psi$ iff $a \vDash \varphi$ and $a \vDash \psi$,

$a \vDash \varphi \vee \psi$ iff $a \vDash \varphi$ or $a \vDash \psi$,

$a \vDash \varphi \to \psi$ iff for any $b \leq a$, if $b \vDash \varphi$, then $b \vDash \psi$.

So, Kripke semantics and SSIL differ in three important respects. First, the semantic structures are different. In both cases, these structures are partially ordered sets but, in contrast with Kripke frames, in any algebra of information states there is the least element and the least upper bound for any two elements. Moreover, in what follows, distributivity will always be required.

Second, valuations are different. In Kripke models, valuations assign any downward closed sets. In SSIL, valuations assign special downward closed sets called ideals.

Third, the semantic clauses for $\bot$ and $\vee$ differ in the two frameworks. While $\bot$ is not assertible in any state of any Kripke model, it is always assertible in the state 0 in SSIL. (And this is the only state in which it is assertible.) In fact, every formula of L is assertible in 0, as can be easily verified by induction. There is no such 'absurd state' in Kripke semantics. However, note that if we added the absurd state to a given Kripke model, it would not have any impact on the assertibility relation $\vDash$ between its original states and formulas.

While in Kripke semantics $\varphi \vee \psi$ is assertible only if φ or ψ is assertible, in SSIL $\varphi \vee \psi$ might be assertible even if neither φ, nor ψ is assertible. The constructive reading of disjunction is in accordance with intuitionistic philosophy. However, this reading seems to be quite unnatural with respect to a relation of assertibility. We typically assert a disjunction when we do not have enough evidence for the assertion of any disjunct. In this connection, note that the presence of the absurd state guarantees the validity of the schema $\varphi \rightarrow (\varphi \vee \psi)$ in SSIL. Suppose that φ is assertible in a. ψ is definitely assertible in 0 and $a + 0 = a$. So, $\varphi \vee \psi$ is assertible in a.

Now, when we have discussed the differences between the two frameworks, let us consider how they are connected to each other. The following concept will serve us as a useful tool for the comparison.

Definition 6 *Let $\mathcal{M}$ be an information model and $\mathcal{K}$ a Kripke model. We say that $\mathcal{M}$ and $\mathcal{K}$ are equivalent if it holds for every formula φ from L that φ is assertible in every state of the model $\mathcal{M}$ with respect to the relation $\Vdash$ if and only if φ is assertible in every state of the model $\mathcal{K}$ with respect to the relation $\vDash$.*

We will describe a way of transforming any given Kripke model into an equivalent information model. Let $\mathcal{F}$ be a Kripke frame and let $d(\mathcal{F})$ be the set of all downward closed sets in $\mathcal{F}$. Relative to $\mathcal{F}$, we define the following distributive algebra of information states:

$$\mathcal{F}^d = \langle d(\mathcal{F}), \cup, \emptyset \rangle.$$

Let $\mathcal{K} = \langle \mathcal{F}, V \rangle$ be a Kripke model. It can be transformed into the following distributive information model:

$$\mathcal{K}^d = \langle \mathcal{F}^d, V^d \rangle,$$

where the valuation V^d is defined as follows:

for any $a \in d(\mathcal{F})$, $a \in V^d(p)$ iff $a \subseteq V(p)$.

It can be easily verified that V^d is indeed a valuation, since $V^d(p)$ is an ideal in $\mathcal{F}^d$. It is the principal ideal generated by $V(p)$. The following result corresponds to theorem 2 in (Punčochář, 2015).

Theorem 1 *Let $\mathcal{K}$ be a Kripke model and let a be a state of the model $\mathcal{K}^d$ (i.e. a downward closed set in $\mathcal{K}$). Then the following holds for any formula φ from L:*

$a \Vdash \varphi$ in $\mathcal{K}^d$ iff for any $b \in a$, $b \vDash \varphi$ in $\mathcal{K}$.

As a consequence, $\mathcal{K}$ and $\mathcal{K}^d$ are equivalent.

So we have described how to transform any given Kripke model into an equivalent distributive information model. Now we will explore how to transform any given distributive information model into an equivalent Kripke model. Since the finite case is of some interest, we will consider the general (finite and infinite) case and the finite case separately.

It can be observed that in the finite case the correspondence between the two semantic frameworks stems from the famous correspondence between finite distributive lattices and finite partially ordered sets. Our exposition of this correspondence is based on (Grätzer, 2011).

Definition 7 *Let $\mathcal{A} = \langle S, +, 0 \rangle$ be a finite distributive algebra of information states and $a \in S$. We say that a is join-irreducible if $a \neq 0$ and the following holds for any $a, b \in S$:*

$$a = b + c \text{ implies } a = b \text{ or } a = c.$$

The set of join-irreducible elements in $\mathcal{A}$ will be denoted as $JI(\mathcal{A})$.

Note that at join-irreducible elements the semantic clause for disjunction

$a \Vdash \varphi \vee \psi$ iff there are b, c such that $b \Vdash \varphi$, $c \Vdash \psi$ and $b + c = a$

reduces to the standard clause of Kripke semantics:

$a \Vdash \varphi \vee \psi$ iff $a \Vdash \varphi$ or $a \Vdash \psi$.

Now we can assign to any finite distributive algebra of information states $\mathcal{A}$ the following partially ordered set:

$$\mathcal{A}^i = \langle JI(\mathcal{A}), \leq^i \rangle,$$

where the ordering $\leq^i$ is the restriction of the ordering $\leq$ in $\mathcal{A}$ to the set $JI(\mathcal{A})$, i.e., the following holds for any $a, b \in JI(\mathcal{A})$:

$a \leq^i b$ iff $a \leq_\mathcal{A} b$.

The famous correspondence between finite distributive lattices and finite partially ordered sets can be expressed in the following way. For any finite distributive algebra of information states $\mathcal{A}$, we have that

(1) $\mathcal{A}$ is isomorphic to $(\mathcal{A}^i)^d$.

A Nonstandard Semantic Framework

Moreover, for any finite Kripke frame $\mathcal{F}$, it holds that

(2) $\mathcal{F}$ is isomorphic to $(\mathcal{F}^d)^i$.

We will build on the fact (1). An important consequence of this fact is that every finite distributive algebra of information states can be represented as an algebra of sets where the zero element is the empty set and the join operation is union.[2] The isomorphism between $\mathcal{A}$ and $(\mathcal{A}^i)^d$ is defined with the help of the following definition.

Definition 8 *Let $\mathcal{A} = \langle S, +, 0 \rangle$ be a finite distributive algebra of information states and $a \in S$. The spectrum of a, denoted as $spec(a)$, is the set of join-irreducible elements that are under a. In symbols:*

$$spec(a) = \{b \in JI(\mathcal{A}); b \leq a\}.$$

The isomorphism h between $\mathcal{A}$ and $(\mathcal{A}^i)^d$ is defined by

$$h(a) = spec(a).$$

The next step is to take into account also the valuations. We will assign to any given finite distributive information model $\mathcal{M} = \langle \mathcal{A}, V \rangle$, the Kripke model

$$\mathcal{M}^i = \langle \mathcal{A}^i, V^i \rangle,$$

where the valuation V^i is defined as follows:

for any $a \in JI(\mathcal{A})$, $a \in V^i(p)$ iff $a \in V(p)$.

Note that valuations are defined in such a way that they respect the isomorphism between $\mathcal{A}$ and $(\mathcal{A}^i)^d$. It holds:

$$a \in V(p) \text{ iff}^3 \; h(a) \subseteq V^i(p) \text{ iff } h(a) \in (V^i)^d(p).$$

As a consequence, for any formula from L, the following holds:

$$a \Vdash \varphi \text{ in } \mathcal{M} \text{ iff } h(a) \Vdash \varphi \text{ in } (\mathcal{M}^i)^d.$$

Now theorem 1 gives us the following conclusion.

---

[2] This can be generalized for every distributive algebra of information states.

[3] The implication from $h(a) \subseteq V^i(p)$ to $a \in V(p)$ follows from the fact, that any element of any finite distributive algebra of information states is identical with the sum of all the elements from its spectrum.

Theorem 2 *Let $\mathcal{M}$ be a finite distributive information model and let a be a state of this model. Then the following holds for any formula φ from L:*

$a \Vdash \varphi$ in $\mathcal{M}$ iff for any $b \in spec(a)$, $b \vDash \varphi$ in $\mathcal{M}^i$.

As a consequence, $\mathcal{M}$ and $\mathcal{M}^i$ are equivalent.

So we have a method of transforming any finite distributive information model into an equivalent Kripke model. A similar transformation is used in standard algebraic semantics. It can be illustrated with this picture:

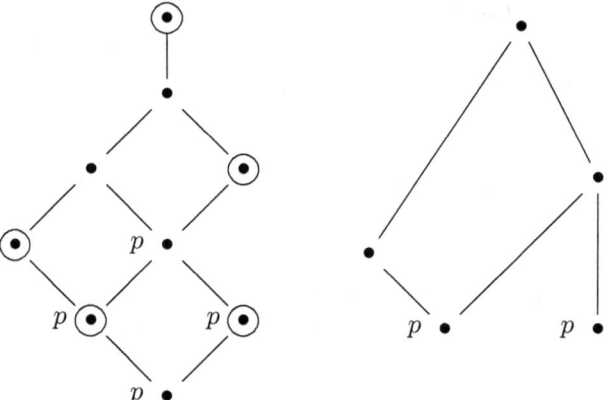

$\mathcal{M}$ and its join-irreducible elements   The Kripke model $\mathcal{M}^i$

Now we want to describe a general method which would work for every distributive algebra of information states. This method is obtained by combining two results. The first one was mentioned at the end of the previous section: the algebra of propositions is the Heyting algebra of ideals. Now it is known how to construct an equivalent Kripke model to any given Heyting algebra enriched with a valuation (in the sense of algebraic semantics). This model is built up from prime filters in the Heyting algebra (see, e.g., Fitting, 1969, chapter 1). So, for a given distributive information model, we can take the lattice of its ideals and construct an equivalent Kripke model from the prime filters in this lattice. For this purpose, let us introduce the following notation. For any algebra of information states $\mathcal{A}$, let $\mathcal{I}(\mathcal{A})$ denote the set of its ideals (ordered by inclusion). For any ideals I, J, let $I \otimes J$ denote the greatest lower bound, and $I \oplus J$ the least upper bound, of $\{I, J\}$ in the lattice of ideals. For distributive algebras of information states, it holds that:

A Nonstandard Semantic Framework

$I \otimes J = I \cap J$,

$I \oplus J = \{a + b; a \in I, b \in J\}$.

Definition 9 *Let $\mathcal{A}$ be a distributive algebra of information states. A nonempty set of ideals $F \subseteq \mathcal{I}(\mathcal{A})$ is called a prime filter in $\mathcal{I}(\mathcal{A})$ if the following two conditions are satisfied for any $I, J \in \mathcal{I}(\mathcal{A})$:*

$I \otimes J \in F$ iff $I \in F$ and $J \in F$.

$I \oplus J \in F$ iff $I \in F$ or $J \in F$.

We say that a prime filter F is proper if $F \neq \mathcal{I}(\mathcal{A})$. The set of all proper prime filters in $\mathcal{I}(\mathcal{A})$ will be denoted as $\mathcal{PFI}(\mathcal{A})$.

We assign to any distributive information model $\mathcal{M} = \langle \mathcal{A}, V \rangle$ the following Kripke model:

$\mathcal{M}^* = \langle \mathcal{A}^*, V^* \rangle$, where $\mathcal{A}^* = \langle \mathcal{PFI}(\mathcal{A}), \supseteq \rangle$,

and the valuation V^* is defined in the following way:

for any $F \in \mathcal{PFI}(\mathcal{A})$, $F \in V^*(p)$ iff $V(p) \in F$.

Now we can apply the famous result according to which the Kripke model constructed in the described way has to be equivalent to the Heyting algebra of ideals and we obtain the following theorem.

Theorem 3 *Let $\mathcal{M}$ be a distributive information model and F a proper prime filter in the lattice of its ideals. Then the following holds for any formula φ from L:*

$F \Vdash \varphi$ in $\mathcal{M}^*$ iff $||\varphi||_{\mathcal{M}} \in F$.

As a consequence, $\mathcal{M}$ and $\mathcal{M}^*$ are equivalent.

It is an open problem whether the ideal completion is necessary, i.e., whether it would be possible to construct an equivalent Kripke model directly from the prime filters in the original algebra of information states instead of in the lattice of its ideals.[4]

---

[4] Let $\mathcal{A}$ be a distributive algebra of information states and F a set of its states. We say that F is a prime filter in $\mathcal{A}$ if (a) F is upward closed; (b) F is downward directed, i.e., if $a \in F$ and $b \in F$, then there is a state $c \in F$ such that $c \leq a$ and $c \leq b$; (c) if $a + b \in F$, then $a \in F$ or $b \in F$.

4 The first-order case

In this section, we will extend the framework to the case of first-order logic. For the sake of simplicity, we will work with a language without functors. So, the basic symbols are brackets, names, predicates of all arities, variables, the constant $\bot$, quantifiers ($\forall, \exists$), and propositional connectives ($\wedge, \vee, \rightarrow$). Well-formed formulas are built up from these symbols in the standard way. Let us denote the resulting language as L^*.

For the sake of simplicity, we will work with constant domains. A first-order information model is defined as a tuple $\mathcal{M} = \langle \mathcal{A}, U, V \rangle$, where $\mathcal{A}$ is a complete lattice, i.e., a partially ordered set (of information states) such that for every set of its elements X, there exists the least upper bound ΣX, U is a nonempty set (the universe of discourse), and V is a function that assigns to every name an element of U, and to every n-ary predicate a function that assigns to every n-tuple of elements from U an ideal in $\mathcal{A}$.

In the algebra of information states $\mathcal{A}$, the least element 0 can be defined as the least upper bound of the empty set $\Sigma \emptyset$. The following notation will also be used: $a + b$ will be the abbreviation for $\Sigma \{a, b\}$ and $a \times b$ the abbreviation for $\Sigma \{c; c \leq a \text{ and } c \leq b\}$.

An evaluation is a function that assigns an element from U to every variable of the language L^*. If e is an evaluation, x a variable and m an element from U, then $e(x/m)$ is the evaluation that assigns m to x and $e(y)$ to every other variable y. Variables and names are called terms. If t is a term, V a valuation and e an evaluation, then $V^e(t)$ is identical with $V(t)$ if t is a name, and with $e(t)$ if t is a variable.

Let us fix a first-order information model $\mathcal{M} = \langle \mathcal{A}, U, V \rangle$. With respect an evaluation e, we can define an assertibility relation $\Vdash_e$ between the formulas from L^* and information states from $\mathcal{A}$. The relation is defined as follows:

$a \Vdash_e \bot$ iff $a = 0$,

$a \Vdash_e Pt_1, \ldots, t_n$ iff $a \in V(P)(V^e(t_1), \ldots, V^e(t_n))$,

$a \Vdash_e \varphi \wedge \psi$ iff $a \Vdash_e \varphi$ and $a \Vdash_e \psi$,

$a \Vdash_e \varphi \vee \psi$ iff there are b, c such that $b \Vdash_e \varphi$, $c \Vdash_e \psi$ and $b + c = a$,

$a \Vdash_e \varphi \rightarrow \psi$ iff for any $b \leq a$, if $b \Vdash_e \varphi$, then $b \Vdash_e \psi$,

$a \Vdash_e \forall x \varphi$ iff for every $m \in U$, $a \Vdash_{e(x/m)} \varphi$,

A Nonstandard Semantic Framework

$a \Vdash_e \exists x \varphi$ iff there is a function g that assigns to every $m \in U$ a state $g(m)$ such that $g(m) \Vdash_{e(x/m)} \varphi$ and $\Sigma\{g(m); m \in U\} = a$.

The semantics starts to behave in a reasonable way when we require that the algebra of information states is infinitely distributive, i.e., that it holds:

$$a \times \Sigma\{b; b \in X\} = \Sigma\{a \times b; b \in X\}.$$

If this condition is satisfied, then it holds for every formula φ from L^* that the set of states in which φ is assertible (with respect to a given evaluation) is an ideal. The semantic framework introduced in this section is a straightforward generalization of the propositional case. A more detailed investigation of the framework is left for future research.

5 Conclusion

To sum up, we have introduced a semantic framework for intuitionistic logic that can serve as an alternative to Kripke semantics and we studied how it is related to Kripke semantics. The proposed framework is denoted as SSIL and its semantic structures are called information models. We have shown that for any Kripke model $\mathcal{K}$, one can build up an equivalent information model from the downward closed sets of states in $\mathcal{K}$ (we worked with a reversed order, so in the standard formulation of Kripke semantics, one would have to take the upward closed sets). Given a (distributive) information model $\mathcal{M}$, one can construct an equivalent Kripke model from the prime filters in the lattice of ideals in $\mathcal{M}$. These transformations show that the framework is related to Kripke semantics in a similar way in which standard algebraic semantics for intuitionistic logic is related to Kripke semantics.

References

Ciardelli, I. (2016). *Questions in Logic*. Doctoral dissertation, University of Amsterdam.
Ciardelli, I., & Roelofsen, F. (2011). Inquisitive Logic. *Journal of Philosophical Logic, 40*, 55–94.
Fine, K. (2014). Truth-maker Semantics for Intuitionistic Logic. *Journal of Philosophical Logic, 43*, 221–246.
Fitting, M. (1969). *Intuitionistic Logic. Model Theory and Forcing*. Amsterdam: North-Holland Publishing Company.

Grätzer, G. (2011). *Lattice Theory: Foundation*. Berlin: Birkhäuser.
Kripke, S. (1965). Semantical Analysis of Intuitionistic Logic I. In M. Dummett & J. N. Crossley (Eds.), *Formal Systems and Recursive Functions* (pp. 92–130). Amsterdam: North-Holland Publishing Company.
Punčochář, V. (2014). A New Semantic Framework for Modal Logic. *Philosophical Alternatives*, *23*, 47–59.
Punčochář, V. (2015). A Generalization of Inquisitive Semantics. *Journal of Philosophical Logic*. (Online first) doi: 10.1007/s10992-015-9379-1
Punčochář, V. (submitted). Algebras of Information States.
Urquhart, A. (1972). Semantics for Relevant Logics. *The Journal of Symbolic Logic*, *37*, 159–169.
Veltman, F. (1984). Data semantics. In J. Groenedijk, T. M. V. Janssen, & M. Stokhof (Eds.), *Truth, Interpretation and Information: Selected Papers from the Third Amsterdam Colloquium* (pp. 43–64). Dordrecht: Foris Publications Holland.
Wansing, H. (1993). Informational Interpretation of Substructural Propositional Logics. *Journal of Logic, Language and Information*, *2*, 285–308.
Yang, F. (2014). *On Extensions and Variants of Dependence Logic*. Doctoral dissertation, University of Helsinki.

Vít Punčochář
Czech Academy of Sciences, Charles University in Prague
The Czech Republic
E-mail: `vit.puncochar@centrum.cz`

Internal Negation and Sortal Quantification

KAREL ŠEBELA

Abstract: This article offers an attempt how to implement the so-called internal negation into the sortal logic. Firstly, I will explain the distinction between internal and external negation and shortly introduce the sortal logic. After that, the problem with implementation of internal negation into sortal logic is shown. Consequently, I will try to present some way out and implement these intuitions into the new definition of internal negation in sortal logic. In the rest of the article I use the traditional square of opposition to uncover various relations between affirmation, external and internal negation.

Keywords: sortal logic, internal negation, external negation, contrariety

1 External vs. internal negation

External negation is the classical boolean negation, internal negation is the negation of a predicate. In natural languages is the internal negation discernible via predicates with prefixes such as "un-", "dis-", "not-", or "non-". In scholastic tradition, internal negation is a "partial" negation of the term, e.g. when x is a non-communist, it is still not negated that (s)he is a man. It is in contrast with the external negation (x is not a communist) which should be true even in the case if x is a stone.

2 Internal negation and contraries (and its history)

The concept of internal negation goes (as usually in logic) back to Aristotle. More specifically, we have to focus on the notions of indefinite name and contrariety. In the book *On Interpretation*, Aristotle introduces so called indefinite names and indefinite verbs, e.g. "not-man", "not-ill". Aristotle called them "indefinite" because their meaning was not positively (definitely) given as in the case of regular ones, but they are still names (or verbs).[1]

Contrariety is one of the four types of opposition, which Aristotle (firstly in *Categories*) distinguishes. Examples of contraries are "healthy" and "ill", "white" and "black". The main difference with respect to contradictories is that in the case of contradictory sentences (like "Socrates is ill" and "Socrates is not ill") one of the sentences is always true and the other false, but in the case of sentences with contrary predicates (like "Socrates is white" and "Socrates is black") both sentences should be false. The point is that sentences with indefinite names behaves like contraries, i.e. both of them should be false, as Aristotle wrote in *Metaphysics*: "everything is equal or not equal, but not everything is equal or unequal, or if it is, it is only within the sphere of that which is receptive of equality".[2] Strictly speaking, indefinite names behaves like immediate contraries. What is it? Boethius, following Aristotle, divides contraries into mediate (black/white, good/bad) and immediate (odd/even, left-hander/right-hander). Names/indefinite names (even/uneven) have no mediate alternatives, so we can label them as immediate contraries.

For scholastic followers of Aristotle, sentences of the form "S is not P" were called negatio negans, sentences of the form "S is non$P$" were called negatio infinitas, whence the label infinite negation for the sentences with indefinite predicates.[3] This kind of negation became the still part of logical tradition, as we can see e.g. in Kant's theory of judgement, where in category of Quantity Kant distinguishes affirmative, negative, and infinite judgements (example is "Soul is immortal").

The era of infinite negation ends with the entry of modern mathematical logic. Frege in his article Negation famously advocates the view that all kinds of negation can be reduced to sentential negation "it is not true, that ...".[4]

3 Internal negation in modern logic

It is difficult to capture the internal negation in classical first-order quantification theory (FQT).

---

[1](Aristotle, 1963).

[2](Aristotle, 1953, 1055b, pp. 10–12).

[3](See Horn, 1989, section 1.1.5). Indefinite negations were in history of logic wrongly labelled as infinite.

[4](See Frege, 1919).

Internal Negation and Sortal Quantification

Consider the sentence

(1) This stone is wise.

It is false. It is representable within classical first-order logic by a formula such as $W(s)$, where $W(x)$ is an open formula representing the property of being wise, s is an individual constant, informally assumed to be denoting the stone in question. The (boolean) negation of the first sentence, namely

(2) This stone is not wise.

Represented by $\neg W(s)$, is true. Here, the open formula $\neg W(x)$ represents the property of not being wise.

However, in a sense, wisdom cannot be attributed to nor denied of a stone. The notion of denying is captured by the Aristotelian concept of term negation. To deny that s has the attribute W is to attribute to s the term negation of W. Importantly, the term negation of W is itself a predicate, not an open formula. For example, the term negation associated with (1) is

(3) This stone is unwise.

Importantly, (3) seems to be false as well and, consequently, law of excluded middle fails for term negation. Thus, (1) and (3) are not contradictories but, rather, they are contraries.

Generally, canonical translation of the "x is not an A" in FQT is $\neg A(x)$. But a canonical translation of the "x is a $not - A$" in FQT is also $\neg A(x)$. One of the reasons is the fact that in the case of internal negation there are some limitations of the universe of discourse. This fact is not (directly) reflected in FQT.

The difference between these two kinds of negation should be explained otherwise. That is the reason for searching the alternatives.

3.1 Some modern formalisations

McCall (1967) outlines a formalisation of a contrary-forming sentential operator within propositional logic. He is criticised by Geach (1969) for formalising the notion of a contrary as an operator, and by Englebretsen (1974) for formalising contraries by a sentential operator. A contrary-forming operator in propositional logic is studied also by Humberstone (2008), see also

Karel Šebela

(Humberstone, 2011). A contrary-forming sentential operator within first-order logic is studied by Wessel (1983), whose approach is set in the context of propositional modal logic by Wojciechowski (1997).

(Sommers, 1967, 1970, 1982) outlines the 'Logic of Terms'—an upgrade of Aristotelian logic, that captures term negation as an operator on predicates.[5]

4 Internal negation and sortal logic

My thesis is that a sortal quantification theory (SQT) could be an appropriate tool to deal with internal negation. The reason is simply that SQT is a form of restricted quantification, which is the case of internal negation, because of the aforementioned limitations of the universe of discourse.

Now, what is SQT? SQT is a version of FQT. The key concept of SQT is—of course—the concept of sortal. The simplest and widely accepted interpretation of sortals is that it gives a criterion for counting the items of that kind, as it is in Cocchiarella's definition of a sortal concept—"a sociogenetically developed cognitive ability or capacity to distinguish, count and collect or classify things".[6] Typical examples of sorts are tigers, cats, tables etc., in short countable items.

Now, SQT introduces so called sortal quantification. The key idea is as follows: consider the sentence

(4) All men are mortal.

In FQT, this sentence is translated as

(4.1) $(\forall x) S(x) \to P(x)$.

Strictly speaking, the reformulated sentence tells us that for every object is true, that if the object is S, then the object is also P. This sounds in a way weird, because the original sentence does not seems to be "about" all objects. In SQT, the original sentence is reformulated as

(4.2) $(\forall x S) x P$.

---

[5]Sections 1,2, and 3 also appear in (Sedlár & Šebela, n.d.).
[6](See Cocchiarella, 1977, p. 441).

Internal Negation and Sortal Quantification

The subformula in brackets means that the universal quantifier quantifies not over all individuals, but over the individuals which falls under S, in short quantifies over all S's. Generally speaking, in SQT the universe of discourse is sorted. But if we would like to introduce the "classical" quantifiers, we could define them as follows:

$$(\forall x)\varphi =_{df} (\forall S)(\forall xS)\varphi$$

$$(\exists x)\varphi =_{df} (\exists S)(\exists xS)\varphi$$

Moreover, in some versions of SQT, sortals are subordinated to another sortals, so we can build a hierarchy of sortals. Every sortal is subordinated to some ultimate sortal, i.e., a sortal which is subordinate to no other sortal. We can write "S^P", where S is a sortal and P is ultimate sortal of S.

Now, if we wish to reformulate the traditional four types of judgements of Aristotelian logic, we could write it as follows:
SaP—$(\forall xS)xP$ SeP—$(\forall xS)\neg xP$ SiP—$(\exists xS)xP$ SoP—$(\exists xS)\neg xP$

5 Problems with internal negation in SQT and a possible way out

At first glance, internal negation seems to be simply a negation of a sortal. But, according to the SQT, negation of a sortal is not a sortal. This principle holds regardless of which interpretation of sortal you take. Thus, according to the proponents of SQT, "is not a cat" is not a sortal (Wiggins), because you cannot count the non-cats since they include dogs, tables, molecules etc. Obviously, here the negation of a sortal means an external negation.

In (Geach, 1962) there is a way out (Geach uses "substantival term" instead of "sortal"): "The negation of a substantival term is never itself a new substantival term. If "the same A" supplies an intelligible criterion of identity, "the same non-A" or "the same thing that is not an A" never of itself does so, though such a criterion may be smuggled in. ("The same non-A" may in context mean "the same B that is not A" where "B" is a substantival term; e.g., "the same nonsmoker" may mean "the same man— or, railway compartment—that is not a smoker".)"[7] My concern here is the "smuggling in", mentioned by Geach. My initial intuition is that internal negation is a not complete characterisation of a sortal. This corresponds to Aristotle's name for expressions like a "not-man"—an indefinite name.

---

[7](See Geach, 1962, p. 64)

We can take it in connection with the classical definition of definition. The classical definition of definition is as follows:

(5) Definitio fit per **genus proximum** et **differentiam specificam**.

If we take an example of this kind of definition

(6) Smoker is a **man** who **smokes tobacco**.

Here "man" is the genus proximum and "smoking tobacco" is the differentiam specificam. If we wish to distinguish the non-smokers, we can say that

(7) Non-smoker is a **man** who does **not smoke tobacco**.

Here we have the same genus—man—but in the place of differentiam specificam there is only a negation of this specification. Consequently, from "x is a non-smoker" we can get an information (if the statement is true), that x is a man and x is not a smoker. So to be a smoker is in a way subordinated to be a man and there is at least one another "kind" (subsortal) of man, not identical with smoker, which is not "positively" stated (and at least one individual falls under such a sortal). Thus, non-smoker—as opposed to "not a smoker"—is a sortal. It gives a criterion for counting the items of that kind.

6 Implementation of internal negation in SQT

Internal negation is thus a "negative" specification of a sortal. In SQT (e.g. in a version in L. Stevenson's *A formal theory of sortal quantification*), we can define the relation of subordination between sortals. Internal negation of a sortal can be then defined in a second-order SQT as an operation on sortals. Informally, it takes a sortal as an argument and returns another sortal as a value.

In more precise terms, we can define internal negation like this: internal negation of sortal S means that there exists sortals S, P, and T such that S and P are subordinated to T and P is different from S. In a formal way

(Def) $\sim S$ is an **internal negation** of S iff $(\exists S)(\exists P)(\exists T)((S \supseteq T) \wedge (P \supseteq T) \wedge (S \cap P = \emptyset))$.

Internal Negation and Sortal Quantification

In this definition, P is a name of a sortal, which is given only via differentiation from S. The specificity of this sortal is that it is defined only via negation of another sortal, so $\sim S$ includes all T's which are not S's. The occurrence of "P" simply means that if there are T's other than S's then there exists at least one sortal subordinated to T which is different from S. $\sim S$ is identical with P.

Analysis of the sentences containing internal negation should be following:

Consider the sentence

(8) John is a non-smoker.

In accordance with (Def), we get

(8.1) $(\forall xS)xP \wedge jP \wedge \neg jS$.

Here "j" stands for John, "S" for smoker and "P" for man. Accordingly, the general sentence

(9) All logicians are non-smokers

could be analysed this way (where "L" stands for logician)—

(9.1) $(\forall xS)xP \wedge (\forall xL)xP \wedge (\forall xL)\neg xS$.

If we would like to simplify this relatively complex expression, we could use the superscript not only for ultimate sortals, but also simply for higher sortals, and then simplify it to

(9.2) $(\forall xL^P)\neg xS^P$.

7 Internal negation and square of opposition

If we wish to analyse the relation between the internal and external negation more deeply, we can use the traditional square of opposition.[8] Relations between the original sentence and its external and internal negation can be captured in it this way:

---

[8]The general idea is taken from the chapter 46 of the first Book of Aristotle's *Prior Analytics*.

Karel Šebela

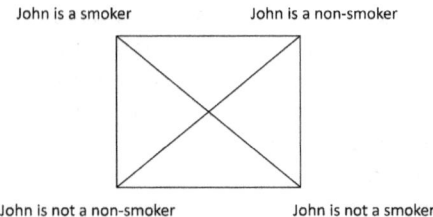

In SQT, we can write

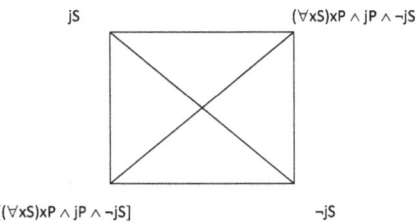

Then, we can interpret sentences like "All S's are non-P's" in SQT as follows: $(\forall xS){\sim}xP \leftrightarrow (\forall xS)\neg xP \wedge (\forall xS)xT \wedge (\forall xP)xT$. "All S's are not non-P's: $(\forall xS)\neg{\sim}xP \leftrightarrow \neg((\forall xS)\neg xP \wedge (\forall xS)xT \wedge (\forall xP)xT)$. These two formulas form a contradictory pair, and since the formulas $(\forall xS)xP$ and $(\forall xS)\neg xP$ form also a contradictory pair, it will be shown that these formulas can also create a kind of a square of opposition, where the traditional relations of contrariety, subalternation and subcontrariety also holds.

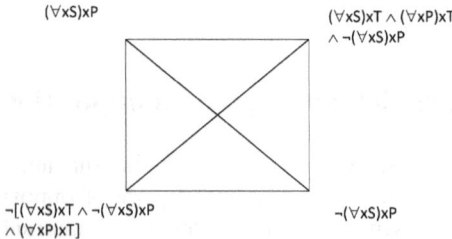

The relation of subalternation between $(\forall xS)xP$ and $(\forall xS)\neg{\sim}xP$ is of a special importance, because the articulation of the latter formula in natural

Internal Negation and Sortal Quantification

language can be easily misinterpreted as a case of simple double negation and so wrongly evaluated as equivalent with the former. Generally, between affirmation and internal negation there is a contrariety, as was required. External negation is subalternated to the internal negation and external negation is in a relation of subcontrariety with the (external) negation of internal negation.

7.1 Validity

Validity of the relations in this square of opposition can be seen better if we substitute A for $(\forall x S) x P$ and B for $(\forall x S) x T \land (\forall x P) x T$. With the help of the previous equivalences, we can then construct a square of opposition with A, $\neg A$ and $\neg A \land B$, $\neg(\neg A \land B)$ as contradictories.

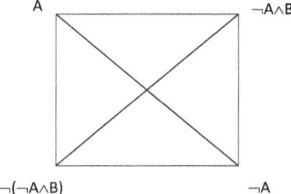

Here we can see the validity of the respective relations clearly.

Contradiction is such that always one of the judgements must be true, and the other false. A and $\neg A$, resp. $\neg A \land B$ and $\neg(\neg A \land B)$ meet these requirements obviously.

Contrariety is such that both judgements (A and $\neg A \land B$) cannot at the same time be true. In our case, if A is true, then $\neg A$ is false, therefore the conjunction $(\neg A \land B)$ is also false. If $(\neg A \land B)$ is true, then both conjuncts are true, so $\neg A$ is true, therefore A is false.

Subcontrariety is such that both judgements can be true, but they cannot both be false. In our case, if $\neg A$ is false, then the first conjunct of $\neg(\neg A \land B)$ is false, so the whole conjuction $(\neg A \land B)$ is false, therefore its negation $\neg(\neg A \land B)$ is true. Similarly, if $\neg(\neg A \land B)$ is false, then $(\neg A \land B)$ is true, so both conjuncts are true, therefore $\neg A$ is true.

Subalternation is such that the inferior judgement is implied by the superior, but not *vice versa*. On the left side we have superior judgement A and inferior $\neg(\neg A \land B)$. If A is true, so $\neg A$ is false, then one part of $(\neg A \land B)$ is false, so $(\neg A \land B)$ is false, therefore its negation $\neg(\neg A \land B)$ is true. On the right side we have superior judgement $\neg A \wedge B$ and inferior $\neg A$. If $\neg A \wedge B$ is true, so both conjuncts are true, therefore $\neg A$ is true.

A similar case should be done if we will take the singular judgements, xS, $\sim xS$, $\neg\sim xS$ and $\neg xS$. Theory of internal negation is thus presented as simply implementable into the SQT.

References

Aristotle. (1953). *Metaphysics*. Oxford: Clarendon Press. (Translated with commentary by W. D. Ross)

Aristotle. (1963). *Categories and De Interpretatione*. Oxford: Oxford University Press. (Translated with notes by J. Ackrill)

Cocchiarella, N. (1977). Sortals, Natural Kinds and Re-dentification. *Logique et Analyse, 20*, 438–474.

Englebretsen, G. (1974). A Note on Contrariety. *Notre Dame Journal of Formal Logic, 15*, 613–614.

Frege, G. (1919). Negation. In M. Beaney (Ed.), *The Frege Reader* (pp. 346–361). Hoboken: Blackwell Publishing.

Freund, M. (2000). A Complete and Consistent Formal System for Sortals. *Studia Logica, 65*, 367–381.

Geach, P. T. (1962). *Reference and Generality*. Ithaca: Cornell University Press.

Geach, P. T. (1969). Contradictories and Contraries. *Analysis, 29*(6), 187–190.

Grandy, R. E. (2014). Sortals. In E. N. Zalta (Ed.), *The Stanford Encyclopedia of Philosophy* (Spring 2014 ed.). Retrieved from `http://plato.stanford.edu/archives/spr2014/entries/sortals`

Horn, L. (1989). *A Natural History of Negation*. Chicago: Chicago University Press.

Humberstone, L. (2008). Contrariety and Subcontrariety: the Anatomy of Negation (with Special Reference to an Example of J.-Y. Béziau). *Theoria, 71*, 241–262.

Humberstone, L. (2011). *The Connectives*. Cambridge, Massachusetts: MIT Press.

McCall, S. (1967). Contrariety. *Notre Dame Journal of Formal Logic, 8*, 121–132.

Pelletier, F. J. (1972). Sortal Quantification and Restricted Quantification. *Philosophical Studies, 23,* 400–404.
Sedlár, I., & Šebela, K. (n.d.). *Term Negation in First-order Logic.*
Sommers, F. (1967). On a Fregean Dogma. In I. Lakatos (Ed.), *Problems in the Philosophy of Mathematics. Proceedings of the International Colloquium in the Philosophy of Science, London, 1965* (Vol. 1, pp. 47–81). Amsterdam: North-Holland Publishing Company.
Sommers, F. (1970). The Calculus of Terms. *Mind, 79,* 1–39.
Sommers, F. (1982). *The Logic of Natural Language.* Oxford: Clarendon Press.
Stevenson, L. (1975). A Formal Theory of Sortal Quantification. *Notre Dame Journal of Formal Logic, 16,* 185–207.
Wallace, J. R. (1965). Sortal Predicates and Quantification. *The Journal of Philosophy, 62,* 8–13.
Wessel, H. (1983). *Logik.* VEB Deutscher Verlag der Wissenschaften.
Wiggins, D. (1980). *Sameness and Substance.* Cambridge, Massachusetts: Harvard University Press.
Wojciechowski, E. (1997). External and Internal Negation in Modal Logic. *Conceptus, 30,* 57–66.

Karel Šebela
Palacký University Olomouc
The Czech Republic
E-mail: `karel.sebela@upol.cz`

Barbourian Temporal Logic

PETR ŠVARNÝ[1]

Abstract: The article presents the first steps in the logical formalization of Barbour's views on time. The models are compared to Branching space-times models and some basic relations of these two approaches are shown.

Keywords: branching, time, space-time, continuations

1 Introduction

Nicolaus Copernicus, amongst others, taught us that we need to question our perspective and claims we take for granted. Some people might cling to their intuitions instead of following Copernicus even after a few hundred years and many astronomical observations and verifications:

> As I contend that our earth is practically flat except for the hills, mountains and valleys, that no such thing as a globe exists, readers may wonder why the sun is not on view all over the world at one time. My answer is as illustrated. No 1 is the position of the sun at mid-day, in June, in England. At the same time it is midnight in New Zealand, and the mountain, hill or horizon as shown at C would easily prevent a person in New Zealand at D from viewing the sun when over England.
> Does the Earth Rotate? NO!, 1919, William Westfield

Nevertheless, we could argue that the most important lesson from the Copernican revolution is not that Earth rotates around the Sun, but that we can take a different, not intuitive, perspective on nature surrounding us and this perspective might help us to understand it better and simplify our future investigations. In the case of planetary movements, Mars is a good example. It has an apparent retrograde motion in the geocentric view, while in the heliocentric view his orbit is a simple ellipse.

---
[1] This output was created within the project 'Proměny koncepce filosofického vnímání', sub-project FF_VG_2015_074 'Rozšíření modelů větvícího se času' solved at Charles University in Prague from the Specific university research in 2015.

Petr Švarný

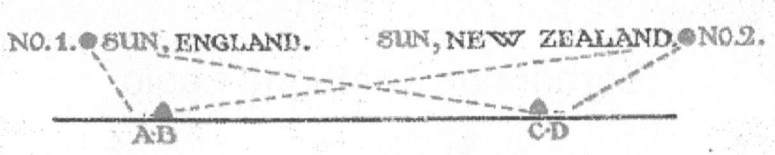

Figure 1: The Westfield Horizon

Isn't it possible that in a similar manner as our existence on the planet Earth influences our perspective of the planetary movements, our existence in time influences our perception of temporal notions? Could there be a similarity between horizons due to Earth's curvature and our perception of the future, also a horizon for our senses? Barbour (2000) argued in his work for a timeless model of the world. However, this perspective might seem at first very unnatural for us, living our daily experiences. We attempt to create a formal temporal logic based on Barbour's ideas. We hope to present a formal basis for the explanation why our temporal statements (and thus existence) still make sense in a timeless universe, in a similar manner as knowledge about planetary movements can help to explain our terrestrial observations of planetary movements.

The article first introduces the main ideas from Barbour's work, followed by the branching space-times formalisms of Belnap and finishing with the Barbour-inspired temporal logic.

2 Barbour's Platonia

A key point for Barbour is the idea presented by Mach:

> It is utterly beyond our power to measure the changes of things... time is an abstraction at which we arrive by means of the changes of things; made because we are not restricted to any one definite measure, all being interconnected. (Barbour, 2000)

Barbour (2000) creates a truly timeless approach to physics which follows the idea that time is merely an abstraction. The second point of inspiration is the Wheeler-DeWitt equation, $\hat{H}(x)|\psi\rangle = 0$, a universe wave function. This wave function should represent all of the universe, with all its possibilities, and according to Barbour we actually are merely living one of the possibilities. As he puts it, the model is a 'modification of Everett's idea: not a many-worlds but a many instants interpretation of quantum mechanics' (Barbour, 2000).

Based on these two basic assumptions, he presents a model of three dimensional snapshots of the universe. These snapshots, sometimes called instants, capture different possible configurations of matter in the universe but no external temporal ordering. All the conclusions about directions of motion (and thus time) are to be found in the intrinsic structure of individual configurations. This intrinsic evidence is later called by Barbour 'time capsules'. A time capsule hence is a part of the universe that suggests in some way the direction of time or evidence for the passage of time, for example geological sediments are a time capsule or particle traces in a cloud chamber. However, the configurations themselves are static and timeless. Nevertheless, thanks to these time capsules, we could arrange the configurations in such a way that they would represent an evolution of the universe (and its possibilities). Barbour also divides the configurations into two 'heaps', possibilities and actualities. Barbour calls this world of configurations, possible or actual, Platonia.

3 Belnap's Our World

The venture point for our logical analysis is Our World. It is with capital letters because we mean Belnap's model for Branching space-times (BST) (Belnap, 1992; Placek & Wroński, 2009). The inspiration is also from physics. BST works with point-events and their sets. We present some of the basic definitions for BST.

Definition 1 (Placek & Wroński, 2009) *The set called Our World OW, is composed of point-events e ordered by $\leq$.*

A set $h \subseteq OW$ is upward-directed iff $\forall e_1, e_2 \in h \; \exists e \in h$ such that $e_1 \leq e$ and $e_2 \leq e$.

A set h is maximal with respect to the above property iff $\forall g \in OW$ such that $h \subset g$, g is not upward-directed.

A subset h of OW is a history iff it is a maximal upward-directed set.

For histories h_1 and h_2, any maximal element in $h_1 \cap h_2$ is called a choice point for h_1 and h_2.

Petr Švarný

Histories are the most important set of point-events in BST. It is meant to capture the familiar notion of possible courses of events. If there a course of events h and there is a inconsistent course of events h' (e.g. the cat is dead or alive), then there is also a point-event $e \in h$ that is inconsistent with a $e' \in h'$. A history can be a very large set of point-events, as we can see from history's definition, depending on the actual BST structure.

Definition 2 (Placek & Wroński, 2009) $\langle OW, \leq \rangle$ *where OW is a non-empty set and $\leq$ is a partial ordering on OW is a structure of BST iff it meets the following requirements:*

1. *The ordering $\leq$ is dense.*

2. *$\leq$ has no maximal elements.*

3. *Every lower bounded chain in OW has an infimum in OW.*

4. *Every upper bounded chain in OW has a supremum in every history that contains it.*

5. *(Prior choice principle) For any lower bounded chain $O \in h_1 - h_2$ there exists a point $e \in OW$ such that e is maximal in $h_1 \cap h_2$ and $\forall e' \in O (e < e')$.*

We can also introduce the language for BST in order to see later, how these semantics can be interpreted in a timeless framework.

Definition 3 (Point fulfils formula—BST) *Let $\mathfrak{M} = \langle OW, \leq, v \rangle$ be a model, with valuation $v : Atoms \to \mathcal{P}(OW)$. For a given event e and history h, st. $e \in h$:*

$\mathfrak{M}, e, h \Vdash p$ iff $e \in v(p)$
$\mathfrak{M}, e, h \Vdash \neg \varphi$ iff not $\mathfrak{M}, e, h \Vdash \varphi$
$\mathfrak{M}, e, h \Vdash \varphi \wedge \psi$ iff $\mathfrak{M}, e, h \Vdash \varphi$ and $\mathfrak{M}, e, h \Vdash \psi$
$\mathfrak{M}, e, h \Vdash F\varphi$ iff there is $e' \in OW$ and $e^* \in h$ s.t.
         $e' \leq e^*$ and $\mathfrak{M}, e', h \Vdash \varphi$
$\mathfrak{M}, e, h \Vdash P\varphi$ iff there is an $e' \in h$ s.t. $e' \leq e$ and $\mathfrak{M}, e', h \Vdash \varphi$
$\mathfrak{M}, e, h \Vdash Sett : \varphi$ iff for all $e' \in h'$, for all h' such that $e \in h'$:
         $M, e', h' \Vdash \varphi$

4 Branching states

How can BST be used to model Barbour's Platonia? At first, we need to realize the difference in scale. While BST speaks about point-events, their ordering, and sets. Barbour worked with configurations which can be looked at as some kind of sets of point-events. If we would simply define configurations as a set of point-events, we could not evade the problems with relativity and observers. In order to be able to prepare these sets, we need to speak instead of point-events of points. Or to be closer to the original idea based on Mach's view, we have to start out by configurations themselves (i.e. the relative positions of all matter, not points on an abstract field). We should also keep in mind the notion of time capsules, especially with respect to semantics.

Definition 4 *We call $\mathcal{P}$ Platonia, the set of all configurations c.*

Configurations are for us a basic building block. However, we assume a deeper structure in them, for example the already mentioned time capsules. This depth, however, can be in our current approach unified into one operator $\Delta(c, c')$, the difference in arrangements (i.e. relative distances, energies, etc.) between two configurations.

Definition 5 *Two configurations $c, c' \in \mathcal{P}$ have a direct transition $c \approx_d c'$ iff $\forall c'' \in \mathcal{P} : \Delta(c, c') \leq \Delta(c, c'')$. There is a transition $c \approx c'$ iff there is a chain of direct transitions $c_1 \approx_d c_2 \approx_d \ldots \approx_d c_n$ such that $c_1 = c$ and $c_n = c'$.*

Therefore if we would look at configurations of two points and have three possible configurations based on only the one dimensional distance of the points: c_1 one meter, c_2 two meters, and c_3 three meters, then there is a direct transition between c_1 and c_2, c_2 and c_3. However, there is not a direct transition between c_1 and c_3, because there exists a configuration whose arrangement is closer to the one of c_1, namely c_2. There still would be a transition between c_1 and c_3.

Definition 6 *Two configurations $c, c' \in \mathcal{P}$ are directly successive $c < c'$ iff $c \approx_d c'$ and $c \in \Psi(c')$. Where $\Psi(c)$ denotes the set of possible preceding configurations based on time capsules from c.*

Definition 7 *A Barbour history h is an direct succession of configurations $c \in \mathcal{P}$.*

Definition 8 *A choice configuration c_c is a configurations $c \in \mathcal{P}$ such that $\exists c_1, c_2 \in \mathcal{P} : c_1 \neq c_2$ and $c_c \in \Psi(c_1) \wedge c_c \in \Psi(c_2)$.*

Definition 9 *Barbour Structure $\mathcal{S}$*

1. *The ordering $<$ is dense.*

2. *The relation $<$ is transitive.*

3. *The relation $<$ is antisymmetric.*

4. *The ordering $<$ has no maximal elements.*

5. *Every lower bounded chain in $\mathcal{P}$ has an infimum in $\mathcal{P}$.*

6. *Every upper bounded chain in $\mathcal{P}$ has a supremum in every history that contains it.*

7. *(PCP) For any lower bounded chain $C \in h_1 - h_2$ there exists a configuration $c \in \mathcal{P}$ such that c is maximal in $h_1 \cap h_2$ and $\forall c' \in C$ $c < c'$.*

The two structures, Barbour's and Belnap's, stay the same at this point. However, notice that in Barbour's idea of configuration ordering there could be a maximal element. A maximal element in this case can represent an ultimate arrangement of matter that does not have any successor (some kind of black hole possibly). This would be also the trivial example of a Barbour structure that is not BST.

We use the language $\mathcal{L}$ with atomic formulas (statements about configurations in the present tense), tense operators F, P, modal operators $Sett$: , $Poss$: and connectives: $\wedge, \vee, \rightarrow, \neg$. The semantic model itself needs only the addition of an interpretation $\mathcal{I} : Atom \rightarrow P(\mathcal{P})$. This interpretation is based on the time capsules of the configurations and their arrangements.

Definition 10 *For the model $\mathcal{M} = \langle \mathcal{S}, \mathcal{I}, \Vdash \rangle$, a c from $\mathcal{P}$ satisfies a formula ψ in language $\mathcal{L}$ iff:*

- *$\psi \in Atom$: $\mathcal{M}, c, h \Vdash \psi$ iff $c \in \mathcal{I}(\psi)$*

- *ψ is $\neg \phi$: $\mathcal{M}, c, h \Vdash \psi$ iff it is not the case that $\mathcal{M}, h \Vdash \phi$*

- *ψ is $\phi \wedge \pi$: $\mathcal{M}, c, h \Vdash \psi$ iff $\mathcal{M}, c, h \Vdash \phi$ and $\mathcal{M}, c, h \Vdash \pi$*

- *ψ is $\phi \vee \pi$: $\mathcal{M}, c, h \Vdash \psi$ iff $\mathcal{M}, c, h \Vdash \phi$ or $\mathcal{M}, c, h \Vdash \pi$*

- ψ is $\phi \to \pi$: $\mathcal{M}, c, h \Vdash \psi$ iff if $\mathcal{M}, c, h \Vdash \phi$ then $\mathcal{M}, c, h \Vdash \pi$

- ψ is $F\phi$: $\mathcal{M}, c, h \Vdash \psi$ iff
  $\exists c' \in \mathcal{P} : c << c'$ and $\exists h' \subset \mathcal{P} : c, c' \in h'$ and $\mathcal{M}, c', h' \Vdash \phi$

- ψ is $P\phi$: $\mathcal{M}, c, h \Vdash \psi$ iff
  $\exists c' \in \mathcal{P} : c' << c$ and $\mathcal{M}, c', h \Vdash \phi$

- ψ is $Sett : \phi$: $\mathcal{M}, c, h \Vdash \psi$ iff
  $\forall h' \subset \mathcal{P} \forall c' \in \mathcal{P} :$ if $c \in h'$ and $(c' < c$ or $c < c')$ then $\mathcal{M}, c', h' \Vdash \phi$

- ψ is $Poss : \phi$: $\mathcal{M}, c, h \Vdash \psi$ iff
  $\mathcal{M}, c, h \Vdash \neg Sett : \neg \phi$

This concludes our brief account of the basics of the Barbour logic. As we see, we can formulate the temporal operators also in this timeless framework using the relation based on time capsules and differences between arrangements.

5 Conclusion

We introduced a preliminary investigation into the a formal system based on Barbour's idea of a timeless universe. Our investigation was merely a starting point for a deeper endeavor as the role of time capsules needs to be investigated further and also the relation to other possible approaches. The article hopes to raise interest in the possibility that timeless frameworks might offer a new perspective on time itself. Hopefully the continuation of such investigations brings us closer to a Copernican explanation of the horizon than to a Westfield-kind of false explanation.

References

Barbour, J. B. (2000). *The End of Time: the Next Revolution in Physics*. Oxford: Oxford University Press.

Belnap, N. (1992). Branching Space-time. *Synthese*, *3*, 385–434.

Placek, T., & Wroński, L. (2009). On Infinite EPR-like Correlations. *Synthese*, *1*, 1–32.

Petr Švarný

Petr Švarný
Department of Logic, Charles University in Prague
The Czech Republic
E-mail: `svarnypetr@gmail.com`

On Paraconsistent Downward Löwenheim-Skolem Theorems

ZACH WEBER[1]

Abstract: A well-known result in inconsistent mathematics is that some paraconsistent arithmetics have *finite models*. This has led to a conjecture that any first order paraconsistent theory has a finite model. The purpose of this note is to investigate the 'finitist' conjecture, to determine how far down the downward Löwenheim-Skolem theorem applies for first order paraconsistent theories. We give methods to determine: (i) which paraconsistent arithmetics have finite models; (ii) which ones must have infinite models, but collapse into classical arithmetic; and (iii) which must have infinite models but are properly paraconsistent.

Keywords: Paraconsistent logic, inconsistent arithmetic, finitism

1 Motivation and background

Any paraconsistent consequence relation $\vdash$ must invalidate two rules: *ex falso quodlibet* or *explosion*,

$$A, \neg A \vdash B$$

and *disjunctive syllogism*

$$A \vee B, \neg A \vdash B$$

This makes dealing with inconsistent theories tenable.

One such projected application (Priest, 2006) would be to *naive proof theory*, which formalizes the 'real' provability predicate $\text{PROV}(x)$ over arithmetic, and thereby should validate the 'S4' laws: theorems prove truths; we can prove that theorems are provable; and proof is closed under modus ponens. This analysis of provability goes back to Gödel (1986). But it leads to

---

[1] Thanks to Guillermo Badia for suggesting the proofs of propositions 1 and 3. Thanks also to audiences at LOGICA 2015 and the March 2015 meeting of the Pukeko Logic Group in Auckland.

'Gödel's paradox': given an (inevitable) sentence G that says $\neg \text{PROV}\ulcorner G\urcorner$, a contradiction follows.[2] Even worse is Löb's theorem: in classical arithmetic, if it were provable that $\text{PROV}\ulcorner A\urcorner$ implies A, then $\text{PROV}(\ulcorner A\urcorner)$, contradicting (on pain of triviality) the soundness principle. A paraconsistent arithmetic, it is hoped, can resolve these issues, either by accepting the Gödel contradiction as true, or by not satisfying Löb's theorem, or both.

One sort of paraconsistent arithmetic is *relevant* arithmetic, developed by Meyer, Mortensen, Routley, Dunn, and Restall, mainly in the strong logic R; see (Dunn, 1980; Meyer & Mortensen, 1984; Restall, 2009). From (Friedman & Meyer, 1992),

> The hope was that $\text{R}^\#$ offered the best of two worlds. On the one hand, its concern for relevance makes $\text{R}^\#$ arguably more reliable than PA. For even if, perish the thought, the Gödel formula is a theorem, there is no way to construct therefrom a proof of $0 = 1 \ldots$ On the other hand, early investigations suggested that $\text{R}^\#$ was as reliable as PA.

Meyer suggested that, indeed, relevant arithmetic can provide its own non-triviality proof (a weak form of soundness) without collapsing into absurdity.

More broadly than relevance, *inconsistent* arithmetics have been developed by Priest in the extensional logic LP, and by Routley in the relevant DK in (Routley, 1977). Priest especially has made great use of a technique in model theory of inconsistent mathematics (Dunn, 1979), the "ultimate downward Löwenheim-Skolem theorem":

Collapsing Lemma *Some paraconsistent arithmetics have finite models.*

The idea is to define an equivalence relation on the natural numbers, and 'collapse' for a new model under which any equivalent numbers are treated as identical. No truths are lost: if A is satisfied in a standard model of arithmetic, A remains satisfied in a collapsed model. Of course, $\neg A$ might also be satisfied in the collapse. Impressed with this from the standpoint of *finitism*, van Bendegem (1993) generalises on the collapsing procedure, leading Bremer (2007) to posit a "paraconsistent downward Löwenheim-Skolem theorem":

---

[2]Proof. From the Gödel sentence, if $\text{PROV}\ulcorner G\urcorner$ then $\neg \text{PROV}\ulcorner G\urcorner$ follows by soundness of proof, so $\neg \text{PROV}\ulcorner G\urcorner$ holds by reductio. And $\neg \text{PROV}\ulcorner G\urcorner$ implies $\text{PROV}\ulcorner \neg \text{PROV}\ulcorner G\urcorner\urcorner$ by the provability of theorems (reflexivity); then $\text{PROV}\ulcorner G\urcorner$ holds by substitution and consequentia mirabilis.

On Paraconsistent Downward Löwenheim-Skolem Theorems

Finitist Conjecture *Any first order paraconsistent theory has a finite model.*

Is the finitist conjecture true? It is by no means an obvious corollary of the collapsing lemma. This note sketches an answer, which clarifies the expressive power of terms like 'finite' in first order paraconsistent theories. One moral is that the expressive power of a paraconsistent language hinges on whether it includes an absurdity constant. Another is that results are relative to the meaning of 'finite' and 'model' in a formal system—which was Skolem's (1967) original point in discussing Löwenheim's results.

2 Defining infinity

Whether or not a theory has a finite model, as opposed to only infinite models, is a matter of the expressive power of the language in which the theory is expressed. Let us review how all standard ways of expressing infinity require *negation*.

According to Dedekind's method, an infinite set has a *proper* subset of the same size. The term 'proper' is doing all the work here; *every* set has a subset of the same cardinality. But to express that X is a proper subset of Y is to say that X is a subset of Y, but something is in Y that is *not* in X: subset X is *non-identical* to Y. Alternatively, we can say that a set is infinite if it contains a subset the size of the natural numbers. This generates infinity because the successor of any $n \in \mathbb{N}$ is *non-identical* to n. In both cases, then, non-identity is doing crucial work (as pointed out by Dunn (1980), citing Meyer).

When negation $\neg$ is paraconsistent and s is arithmetic successor, sentences that express that every number has a non-identical successor, like

$$\forall x \exists y (sx = y \land x \neq y)$$

can be satisfied by a *non-self-identical number*, some

$$k = s(k) \neq s(k) \neq k$$

This is why some paraconsistent arithmetics have finite models. Finitely many non-self-identical objects simulate infinitely many objects.

For a theory of the natural numbers to force infinite models will require more negation. An absurdity constant $\bot$ in the language is presumed to satisfy

$$\bot \to A$$

where '$\to$' is here and throughout a relevant conditional (and so satisfies modus ponens). This gives a sort of negation, $A \to \bot$. This negation is explosive

$$A, A \to \bot \vdash B$$

But it is not exhaustive, since it is not the case that, for all A, either A or else A is absurd. A good candidate for arithmetic absurdity is $\bot := 0 = 1$. If a theory expresses infinity with $A \to \bot$ negation, then we will see that models of it *cannot* be finite.

3 Which paraconsistent arithmetics have finite models?

Arithmetic in the logic LP takes the axioms of classical Peano Arithmetic, phrased with material conditional. Since modus ponens for the material conditional is really just disjunctive syllogism, LP has no detachable implication connective. This makes LP arithmetic easy to satisfy with all sorts of models, e.g. LP has finite collapse models. Because it is so weak, though, LP has no potential for mathematical projects. Notably, one could even add $\bot$ to the language of LP and still have finite satisfaction, because $\bot \supset A$ does not detach. *Any* LP theory has a finite model.

Relevant arithmetic $R^\#$ is a first order axiomatic theory like classical Peano Arithmetic, but phrased with (detachable) relevant $\to$ from R, in the language $\mathcal{L}_R = \{\vee, \neg, \exists, \to, s, 0, +\}$:

RI $0 \neq sx$
RII $x = y \to sx = sy$
RIII $sx = sy \to x = y$
RIV $x + 0 = x$
         $x + sy = s(x + y)$
RV $x \times 0 = 0$
         $x \times sy = (x \times y) + x$
RVI   if $\vdash A(0)$ and $\vdash \forall x(A(x) \to A(sx))$ then $\vdash \forall x Ax$

These axioms may be satisfied by a finite model, like $\mathbb{Z}$ mod 2, as shown by Meyer and Mortensen (1984). In fact, Friedman and Meyer produce a model of relevant arithmetic in the complex ring $\mathbb{C}$, which invalidates some theorems of PA. So $R^\#$ may have finite models, and $R^\#$ is not PA-complete.

4 Which can only have infinite models?

It is a little surprising that $R^{\#}$ is so weak. After all, it is a very strong relevant logic. It leads us to ask which paraconsistent arithmetics, if any, can only have infinite models.

To answer this, let's look at a simple theory that forms the 'infinitary core' of arithmetic. With a primitive order relation $<$, *proto-arithmetic* has axioms

$$\text{PI} \quad x < x \to \bot$$
$$\text{PII} \quad x < y \land y < z \to x < z$$
$$\text{PIII} \quad \forall x \exists y (x < y)$$

that define a strict pre-order with no right endpoint. Notice immediately that this means $(x < y \land y < x) \to \bot$, so $<$ cannot be symmetric. The $\to$ can be from as weak a logic as relevant B.

Proposition 1 (Badia) *Proto-arithmetic has only infinite models.*

Proof. Generate a sequence from an arbitrary starting point $a_0, a_1, a_2, \ldots$ in which no two members are identical. The theory, being interpretable in classical PA, has a model with a non-empty domain D. Let $a_0 \in D$. By PIII, $a_0 < a_1$ for some $a_1 \in D$. If $a_0 = a_1$ then $a_0 < a_0$, which by PI is absurd; so a_0 does not equal a_1, on pain of triviality. Then there is some a_2 such that $a_1 < a_2$. Again, a_1 is not identical to a_2. By PII, it is also not the case that $a_0 = a_2$. And so forth. □

A $<$ relation like this is easily definable as $x < y := \exists z(x + sz = y)$. So looking again at $R^{\#}$, then, it seems clear that we could eliminate its finite models by rephrasing axiom RI as

$$\text{RI}' \quad s(x) = 0 \to \bot$$

Unlike $\neg$, the $\bot$ constant demands quantifiable difference (up to triviality). As we saw already with LP, though, $\bot$ alone is not sufficient for infinity. Arithmetic with a weaker version of RI, like

$$\text{RI}'' \quad s(0) = 0 \to \bot$$

has finite models (Friedman & Meyer, 1992). This is not so surprising, since if $\bot$ is defined as $0 = 1$ then RI'' just says $0 = 1 \to 0 = 1$. Absurdity $\bot$ must be put in the right place to do its work, so to speak.

Now we know some minimal conditions for a theory to have an infinite model; it should contain proto-arithmetic. In the other direction, how much 'infinity' can a theory express before it is simply classical arithmetic again?

5 Which ones are (not) just classical?

Let us turn our attention to a fragment of Robinson Arithmetic, Q, formulated relevantly in R:

$$\text{QI} \quad sx \neq 0$$
$$\text{QII} \quad x \neq 0 \to \exists y\, x = sy$$
$$\text{QIII} \quad x = y \to (A(x) \to A(y))$$

The negation here is paraconsistent $\neg$. As Dunn shows, even without $\bot$ in the formulation, this has only infinite models! But the initial surprise wears off when we see the reason.

Proposition 2 (Dunn) *Relevant Robinson Arithmetic is just classical PA.*

Proof. Dunn shows that, from QII, $1 \neq 0 \to 0 = 0$. Then $0 \neq 0 \to 0 = 1$ by contraposition. Meanwhile, $A \to (0 = 0 \to A)$, from QIII and permutation. So $A \to (\neg A \to 0 \neq 0)$ by contraposition, and then

$$A \to (\neg A \to 0 = 1)$$

But $0 = 1 \to B$ is provable for all B, by some clever use of axiom QII. So the apparent weakness of axiom QI is only apparent. This fragment of Q validates explosion and therefore classical logic. □

Some of Dunn's results seem to have more to do with the peculiarities of Q than anything about paraconsistency, e.g. he finds that dropping 0 and starting at 1 does *not* collapse back into classical logic. But what seems to be happening in the above proof is that the phrasing of QII has built in disjunctive syllogism to the more usual expression

$$\text{QIV} \quad x = 0 \lor \exists y (x = sy)$$

That is, axiom QII would follow from QIV plus $(x = 0 \land x \neq 0) \to \bot$.

So we know that there are arithmetics that require infinite models. For all we know, though, we've just been looking at bits of classical PA with notational variants. We want to see that some infinitary arithmetics are *properly* sub-classical.

Proposition 3 (Badia) *Proto-arithmetic is paraconsistent.*

Proof. The proof uses classical model theory, showing the existence of a model of proto-arithmetic that non-trivially satisfies some contradiction. Let $\mathcal{M} = \langle W, D, \star, R \rangle$ be a quantificational Routley-Meyer frame. Worlds $X \in W$ are said to satisfy formulae, written $X \Vdash A$, set D is a domain of quantification, and the operation $\star : W \longrightarrow W$ is such that $X \Vdash \neg A$ iff $X^\star \nVdash A$. Relation R is a three-place relation on worlds. See (Restall, 2009).

For our target model of PI − PIII, the set of worlds is a pair $W = \{N, M\}$, the domain is the natural numbers, $D = \mathbb{N}$, the ternary accessibility relation[3] is $R = \{(N,N,N), (M,M,M), (N,M,M), (M,N,N), (M,N,M), (N,M,N), (M,M,N)\}$ and the two worlds are star-mates,

$$N^\star = M,$$
$$M^\star = N$$

Let the denotation of $<$ at N be the usual ordering on $\mathbb{N}$, while $<$ at M is the same, minus $\langle 1, 2 \rangle$. That is, $M \nVdash 1 < 2$. Therefore $N \Vdash \neg(1 < 2)$ because it is $M^\star$, but $N \Vdash 1 < 2$ by construction. Therefore N satisfies a contradiction, but still not everything, e.g. $N \nVdash \exists x x < x$. So proto-arithmetic is not classical, as required. A fortiori for R$^\#$. □

6 Coda: infinity in the metatheory

To summarize our downward Löwenheim-Skolem facts in terms of language, *any* LP theory in language $\mathcal{L}_{\mathsf{LP}}^\perp = \{\vee, \neg, \exists, s, 0, +, \perp\}$ has a finite model. For a relevant language $\mathcal{L}_{\mathsf{R}}^\perp = \{\vee, \neg, \exists, \rightarrow, s, 0, +, \perp\}$ (decisively with absurdity), the floor is ω. One can remove either of $\perp$ or $\rightarrow$ and return to finite model satisfaction. So we have answered the finitist conjecture in the negative, and along the way identified conditions that make paraconsistent arithmetics strong enough to need infinite models, but weak enough not to be classical PA.

A departing thought, though. The foregoing arguments are conducted in classical metatheory. The serious paraconsistentist could still ask: do any non-classical arithmetics have only infinite models *according to classical model theory*, but finite models *according to a paraconsistent metatheory*? Suppose for instance that

Inf $\quad \exists x (\exists y \, y \in x \wedge \forall y (y \in x \rightarrow y \cup \{y\} \in x))$

---
[3] As verified by G. Badia and A. Tedder using Prover9.

is true in some paraconsistent set theory. This looks to be an axiom of infinity for von Neumann ordinals. But suppose that the reason Inf is true is because of the existence of some fixed point

$$\mathcal{I} = \{x : x = \mathcal{I}\}$$

granted by naive comprehension. This is a set identical to its own singleton. It makes Inf true, but still looks to be no bigger than a singleton. Now, this theory might say something like 'Proto-arithmetic has only infinite models'— but that claim is *itself* satisfiable by a *finite* model. Or at least, the model is finite from a classical perspective...

As some of the open questions about inconsistent mathematics become classically settled, the real (and open) non-classical work can begin.

References

Bremer, M. (2007). Varieties of Finitism. *Metaphysica*, *8*, 131–148.
Dunn, J. M. (1979). A Theorem in 3-valued Model Theory with Connections to Number Theory, Type Theory, and Relevant Logic. *Studia Logica*, *38*, 149–169.
Dunn, J. M. (1980). Relevant Robinson's Arithmetic. *Studia Logica*, *38*, 407–418.
Friedman, H., & Meyer, R. K. (1992). Whither Relevant Arithmetic? *Journal of Symbolic Logic*, *57*, 824–831.
Gödel, K. (1986). An Interpretation of the Intuitionistic Propositional Calculus. In S. Feferman (Ed.), *Kurt Gödel Collected Works* (Vol. 1, pp. 300–302). Oxford: Oxford University Press.
Meyer, R. K., & Mortensen, C. (1984). Inconsistent Models for Relevant Arithmetics. *Journal of Symbolic Logic*, *49*, 917–929.
Priest, G. (2006). *In Contradiction*. Oxford: Oxford University Press.
Restall, G. (2009). Models for Substructural Arithmetics. In M. Bílková (Ed.), *Miscellanea Logica* (pp. 1–20). Prague: Charles University.
Routley, R. (1977). Ultralogic as Universal? *Relevant Logic Newsletter*, *2*, 51–89.
Skolem, T. (1967). Some Remarks on Axiomatized Set Theory. In J. van Heijenoort (Ed.), *From Frege to Gödel*. Cambridge, Massachusetts: Harvard University Press.

On Paraconsistent Downward Löwenheim-Skolem Theorems

van Bendegem, J. P. (1993). Strict Yet Rich Finitism. In Z. W. Wolkowski (Ed.), *First International Symposium on Gödel's Theorems* (pp. 61–79). Singapore: World Scientific Press.

Zach Weber
University of Otago
New Zealand
E-mail: `zach.weber@otago.ac.nz`

www.ingramcontent.com/pod-product-compliance
Lightning Source LLC
Chambersburg PA
CBHW070736160426
43192CB00009B/1461